20
Topics in Medicinal Chemistry

Aims and Scope

Drug research requires interdisciplinary team-work at the interface between chemistry, biology and medicine. Therefore, the new topic-related series Topics in Medicinal Chemistry will cover all relevant aspects of drug research, e.g. pathobiochemistry of diseases, identification and validation of (emerging) drug targets, structural biology, drugability of targets, drug design approaches, chemogenomics, synthetic chemistry including combinatorial methods, bioorganic chemistry, natural compounds, high-throughput screening, pharmacological in vitro and in vivo investigations, drug-receptor interactions on the molecular level, structure-activity relationships, drug absorption, distribution, metabolism, elimination, toxicology and pharmacogenomics.

In general, special volumes are edited by well known guest editors.

In references Topics in Medicinal Chemistry is abbreviated Top Med Chem and is cited as a journal.

More information about this series at http://www.springer.com/series/7355

Dipanjan Pan
Editor

Personalized Medicine with a Nanochemistry Twist

Nanomedicine

With contributions by

F.B. Bombelli · S. Braswell · J. Caffarini · H.-H. Chang · E.A. Daza · D. Di Silvio · D. Frankowski · J. George · D. Goatz · M. Gryka · M. Imgruet · N. Kelleher · S. Kim · N. Kolmodin · C.C. Konopka · M. Kumar · J. Kus · R. Lake · M. Mazurek · M. Modak · A. Nandyala · N. Olsen · F. Ostadhossein · D. Pan · D. Patel · A. Schwartz-Duval · B. Seadler · V. Sherwood · S. Slania · O. Sonoiki · S. Venkataraman · S. Wang · R.C. Yada · A. Zimmer

Springer

Editor
Dipanjan Pan
Department of Bioengineering
University of Illinois at Urbana Champain
Urbana, Illinois
USA

ISSN 1862-2461 ISSN 1862-247X (electronic)
Topics in Medicinal Chemistry
ISBN 978-3-319-33544-5 ISBN 978-3-319-33546-9 (eBook)
DOI 10.1007/978-3-319-33546-9

Library of Congress Control Number: 2016938785

Printed on acid-free paper

This Springer imprint is published by Springer Nature
The registered company is Springer International Publishing AG Switzerland

Preface

Advances in nanotechnology have allowed for potentially earlier identification and treatment of disease pathologies that are of cellular and molecular origin. These enable us to detect biological signatures even before the actual manifestation of a disease, leading to a medicine that is 'pre-emptive' in nature. This approach is a paradigm shift from conventional medicine, which is more 'symptomatic' in character. At the nanoscale, unique properties emerge enabling us to deliver higher local concentrations of cytotoxic drug with minimal systemic concentrations. Current nanoparticle-based treatments are capable of combining modality-specific imaging contrast with high drug payload and large surface area targeting ligands for an advanced multipurpose therapy agent. The combinatory correlation between treatment and localization of disease models is exclusively exploited in the nanomedicine field. 'Theragonism' or 'theranostic' is a relatively new term which defines that multifunctional nanoparticle for personalized medicines as the imaging contrast provided allows tracking the efficacy of the therapy throughout the application. This personalization with concurrent monitoring of medical treatment becomes especially critical when considering diseases that are largely heterogeneous in nature such as cancer, whose current treatments are associated with emaciation and suffering, almost as highly as the disease.

This volume of *Topics in Medicinal Chemistry* is dedicated to discuss the current trend of the next-generation personalized medicine where we mainly focus on the role of nano-architectures and how defined chemistry helps to tune their functional properties for optimum performance in a biological system. For this thematic issue, we have invited the leading experts in this field to contribute. The special issue is comprised of seven review chapters, which includes one introductory chapter. The articles illustrate a presentation of the advancements related to the field of gene delivery, which we hope will stimulate designing better carriers and tune these technologies for basic, translational and clinical applications.

In the introductory article, Dr. Dipanjan Pan at the University of Illinois at Urbana-Champaign highlights the importance of 'nano' in personalized medicine. We briefly learn about the present status of this field in terms of their clinical translation. In another lead article, Dr. Francesca Baldelli-Bombelli at the Politecnico di Milano and her collaborators highlight the importance of understanding the physico-chemical behaviour of the nanoparticles. This article reports on the state-of-art techniques for the physico-chemical characterization of nanoparticle-protein complexes in the biological environment with particular emphasis on their impact on the efficiency and safety of a new generation of nanomedicines. In a following article, Dr. Dipanjan Pan at the University of Illinois at Urbana-Champaign and his co-workers highlight the biological barriers faced by nanomedicines for effective targeting and delivery in vivo. Dr. Srinivasan Venkataraman at the Institute of Materials Research and Engineering in Singapore in his review article introduces the readers to the fascinating world of defined supramolecular assemblies derived from novel polymers for biomedical application. In the following two chapters, Dr. Dipanjan Pan and his colleagues at the University of Illinois at Urbana-Champaign review the present status of multimodal imaging and theranostic and nano-enabled delivery of intracellular therapeutics. In the concluding chapter, Dr. Dipanjan Pan at the University of Illinois at Urbana-Champaign critically reviews the status of the field and its potential to reach the clinic.

Urbana, IL, USA Dipanjan Pan

Contents

Top Med Chem (2016) 20: 1–28
DOI: 10.1007/7355_2014_68

Published online: 20 January 2015

Barriers in Nanomedicine: The Importance of Defined Chemistry and Engineering Approaches for Clinical Translation

Huei-Huei Chang, Shaneen Braswell, Jonathan George, Mark Gryka, Sumin Kim, Nicolas Kolmodin, Manu Kumar, Benjamin Seadler, Oluwayemisi Sonoiki, and Dipanjan Pan

Abstract The multidisciplinary research of nanomedicine unites the unique prospective of nanotechnology with biology and medicine. A myriad of technological advancements has been made over the past two decades demonstrating the high impending growth of this field for clinical translation. In tandem, the advancements

---

H.-H. Chang and S. Braswell
Department of Chemistry, University of Illinois at Urbana Champaign, Urbana, IL, USA

Department of Bioengineering, University of Illinois at Urbana Champaign, Urbana, IL, USA

J. George
College of Engineering, Department of Nuclear, Plasma and Radiological Engineering, University of Illinois at Urbana Champaign, Urbana, IL, USA

M. Gryka
Department of Bioengineering, University of Illinois at Urbana Champaign, Urbana, IL, USA

Beckman Institute for Science and Technology, Urbana, IL, USA

S. Kim
College of Engineering, Department of Bioengineering, University of Illinois at Urbana Champaign, Urbana, IL, USA

N. Kolmodin
College of Engineering, Department of Chemical and Biomolecular Engineering, University of Illinois at Urbana Champaign, Urbana, IL, USA

M. Kumar, B. Seadler, and O. Sonoiki
College of Engineering, Department of Bioengineering, University of Illinois at Urbana Champaign, Urbana, IL, USA

D. Pan (✉)
Department of Bioengineering, University of Illinois at Urbana Champaign, Urbana, IL, USA

Beckman Institute for Science and Technology, Urbana, IL, USA

Carle Foundation Hospital, Urbana, IL, USA
e-mail: dipanjan@illinois.edu

in chemistry, molecular biology, and engineering have molded this field emphasizing the early detection and treatment of diseases at the molecular and cellular level. Myriads of nanomedicine platforms have been proposed and developed and tested in laboratories and in preclinical models. However, very few have been translated to clinical trials. It is, therefore, a critical issue to recognize the factors affecting their eventual application in human. Towards this aim, we critically review our present understanding of the biological and biophysical obstacles encountered by the nanoagents, which we hope will promote the development of nanotechnologies in terms of future translational and clinical applications.

Keywords Biological barriers, Nanomedicine, Nanotoxicology, Translational research

Contents

Abbreviations

BBB	Blood-brain barrier
BLM	Bilayer lipid membrane
DCS	Differential centrifugal sedimentation
ECM	Extracellular matrix
ENM	Engineered nanomaterial
EPR	Enhanced permeability retention
HIFU	High intensity focused ultrasound
i.v.	Intravenous
k_{off}	Dissociation rate constant
MPS	Mononuclear phagocyte system
MS	Mass spectrometry
NIPAM	*N*-Isopropylacrylamide
NLS	Nuclear localization signal
NMR	Nuclear magnetic resonance
NP	Nanoparticle(s)
PC	Protein corona
PEG	Polyethylene glycol
PS	Polystyrene

QCM	Quartz-crystal microbalance
RES	Reticuloendothelial system
SWCNT	Single-walled carbon nanotubes
Tf	Transferrin
TfR	Transferrin receptor

1 Introduction

Nanomedicine is a multidisciplinary area of research which unites the unprecedented potential of nanotechnology with medicine. A myriad of technological advancements has been made over the past two decades demonstrating the high impending growth of this field for clinical translation [1–4]. Concomitantly, the advancements in chemistry, molecular biology, and engineering have shaped this field towards the early detection and treatment of diseases at the molecular and cellular level [4–9]. Hence, innumera-progress has been made to design defined nanostructures for performing multiple functions, e.g., imaging and therapy. Several hundred companies are in the process of developing nanomedicine platforms [10–12]. Many of these platforms have been tested in laboratories and in preclinical models. However, very few have been translated to clinical trials. It is therefore a critical factor to understand what prevents their further development from finding the eventual application in human.

In this chapter, the readers are introduced to the concept of "theranostics" and the challenges created by the innumerous biological barriers that are responsible for averting the successful clinical translation of these agents. The term "theranostics" was coined to define translational research covering personalized medicine and to express the combinatory approach with diagnostic and therapeutic capabilities of a single agent. There is a great importance placed on nanoparticle (NP) composition, size, shape, and surface properties that dictate the in vivo characteristics and closely interlink with their bio-distributive nature, tissue accumulation, and cellular uptake [13–15]. Despite these challenges, selected nanoparticles have already been approved, while some are undergoing preclinical and clinical evaluation. A little over 200 companies are in the process of developing nanomedicine platforms. The examples of such platforms include liposomes, polymeric micelles, dendrimers, quantum dots (Q-dots), gold nanoparticles (AuNPs), titanium oxide NPs (TiO_2 NPs), silica nanoparticles (Silica NPs), etc. (Table 1). Interestingly, a majority of these platforms are dominated by liposomes and polymer-drug conjugates.

One of the earliest examples of "nano" therapy is Doxil (Janssen Biotech)—a polyethylene glycol-coated, doxorubicin-encapsulated liposomal suspension approved for clinical use. The encapsulation of the drug in PEGylated lipid vesicles enhanced their circulatory half-life and improved drug accretion in the tumor tissue. Doxil was approved by the FDA in 1995 to treat an AIDS-associated cancer. Although the drug induced fewer side effects in comparison to its active chemotherapeutic ingredient, doxorubicin (DOX), the patients' survival rates compared with parent doxorubicin were not statistically improved. Doxil represents the first generation of excipient nanomedicine agent sought to improve drug uptake and reduce toxicity.

Table 1 List of globally approved nanoparticles

Product	Nanoplatform/agent	Indication	Status	Company
Doxil	PEGylated liposome/ doxorubicin hydrochloride	Ovarian cancer	Approved 11/17/1995 FDA50718	Ortho Biotech (acquired by JNJ)
Myocet	Non-PEGylated liposomal doxorubicin nanomedicine	Metastatic breast cancer	Approved in Europe and Canada, in combination with cyclophosphamide	Sopherion Therapeutics, I,L,C in North America and Cephalon, Inc., in Europe
DaunoXome	Lipid encapsulation of daunorubicin	First-line treatment for patients with advanced HIV-associated Kaposi's sarcoma	Approved in the USA	Galen Ltd.
ThermoDox	Heat-activated liposomal encapsulation of doxorubicin	Breast cancer, primary liver cancer	Received Fast Track Designation, approval expected by 2013	Celsion
Abraxane	Nanoparticulate albumin/paclitaxel	Various cancers	Approved 1/7/2005 FDA21660	Celgene
Rexin-G	Targeting protein tagged phospholipid/ micro RNA 122	Sarcoma, osteosarcoma, pancreatic cancer, and other solid tumors	Fully approved in Philippine Phase II/III (Fast Track Designation, Orphan Drug Status Acquired in the USA	Epeius Biotechnologies Corp.
Oncaspar	PEGylated asparaginase	Acute lymphoblastic leukemia	Approved 24/06/2006	Enzon Pharmaceuticals, Inc.
Resovist	Iron oxide nanoparticles coated with carboxydextran	Liver/spleen lesion imaging liver/ spleen	In 2001 m approved for the European market	Bayer Schering Pharma AG
Feridex	Iron oxide nanoparticles coated with dextran	Lesion imaging	Approved by USFDA in 1996	Berlex Laboratories
Endorem	Iron oxide nanoparticles coated with dextran	Liver/spleen lesion imaging	Approved in Europe	Guernet

Furthermore, liposomal formulations fell under scrutiny due their low drug entrapment efficiency, and subsequently, polymeric nanoparticles were proposed as alternative drug carriers. Polymeric nanoparticles presented as an attractive alternative due to their well-defined structure and flexibility for structural tweaking. The use of polyethylene glycol (PEG) as a polymeric drug carrier was first introduced in the early 1990s. Advantages of using PEG include increased plasma stability, enhanced solubility of an insoluble drug, and reduced immune response by its stealth action. Abraxane is one of the earliest successful examples of polymer-bound therapeutics paclitaxel, used for the second-line treatment of breast cancer patients. The average hydrodynamic diameter of Abraxane is 130 nm. It was proposed that the agent can be taken up by the "leaky" tumor vasculatures and in part through the transendothelial transport mechanisms via the albumin-binding protein gp-60 (a 60-kDa sialoglycoprotein). This well-studied mechanism is the enhanced permeability and retention (EPR) effect. EPR broadly relates to the mechanism by which passively targeted nanoparticles are taken up by the tumor vasculature.

There are more than fifty nano-formulations being investigated at different levels of clinical development (Table 2). The majority of these agents depend on passive targeting approaches [16, 17]. Passively targeted nanoparticles are not tissue or molecularly targeted, and they do not depend on "active" homing process. A few examples of these promising agents studied by National Cancer Institute (NCI) Alliance members are discussed below.

AuroLase, a silica-coated gold nanoshell particle, was developed by the investigators at Rice University. These particles preferentially accumulate in cancer lesions in a size-dependent manner. AuroLase is being evaluated for the promising effect of photothermal ablation of tumors with higher therapeutic efficacy and markedly diminished side effects [18]. Sofie Biosciences is developing a [18F]-FAC (1-(2′-deoxy-2′-[18F]fluoroarabinofuranosyl) cytosine) family of PET imaging agents for clinical use. These agents are geared for well-known chemotherapeutics, e.g., gemcitabine, cytarabine, fludarabine, etc., for treating metastatic breast, non-small cell lung, ovarian, and pancreatic cancers, as well as leukemia and lymphomas (http://nano.cancer.gov/action/programs/caltech/).

Calando Pharmaceuticals is conducting clinical trials of a cyclodextrin-based polymeric nanoparticle platform that encapsulates a small-interfering RNA (siRNA). An open-label, dose-escalating trial of their candidate agent was directed to understand the safety of this drug in patients resistant to other chemotherapies [19].

Cerulean Pharma, Inc., is developing a conventional chemotherapeutic (camptothecin/CPT) conjugated with the abovementioned polymeric nanoparticles (CRLX101). An open-label, dose-escalation study of CRLX101 (previously named IT-101) is ongoing for solid tumor malignancies [20].

Drs. Gregory Lanza and Samuel A. Wickline initiated a clinical trial to study a MRI contrast agent based on lipid-stabilized perfluorocarbon (PFC) nanoparticles that bind to the $\alpha_v\beta_3$-intregrin found on the surface of the angiogenic blood vessels associated with early tumor development [21]. Earlier, ligand-directed perfluorocarbon nanoparticles were found to be an effective acoustic contrast agent and subsequently helped to expand this platform technology to include magnetic

Table 2 Nanoparticle cancer therapeutics undergoing clinical investigation

Product/ agent	Nanoplatform	Indication	Status	Company
Cydosert	Cyclodextrin nanoparticles (cyclo-dextrin NP/SiRNA)	Solid tumors	Phase I	Insert Therapeutics (now Calando Pharmaceuticals)
CRLX101	Cyclodextrin NPs/Camptothecin	Various cancers	Phase II	Cerulean Pharma
S-CKD602	PEGylated liposomal CKD602 (topoisom-erase inhibitor)	Various cancers	Phase I/II	Alza Corporation
CPX-1	Liposomal irinotecan	Colorectal cancer	Phase II	Celator Pharmaceuticals
CPX-351	Liposomal cytarabine and daunorubicin	Acute myeloid leukemia	Phase I	Celator Pharmaceuticals
LE-SN38	Liposomal SN38	Colorectal cancer	Phase II	Neopharm
INGN-401	Liposomal/FUSl	Lung cancer	Phase I	Introgen
NC-6004	Polymeric nanoparticle (PEG-polyaspartate) formulation of cisplatin	Various cancers	Phase I	NanoCarrier Co.
NK-105	Polymeric nanoparticle (PEG-polyaspartate) formulation of paclitaxel	Various cancers	Phase II	Nippon Kayaku Co. Ltd.
NK-911	Polymeric nanoparticle (PEG-polyaspartate) formulation of doxorubicin	Various cancers	Phase I	Nippon Kayaku Co. Ltd.
NK-012	Polymeric micelle of SN-38	Various cancers	Phase II	Nippon Kayaku Co. Ltd.
SP1049C	Glycoprotein of doxorubicin	Various cancers	Phase II	Supratek Pharma Inc.
SPI-077	PEGylated liposomal cisplatin	Head/neck and lung cancer	Phase II	Alza Corporation
ALN-VSP	Lipid nanoparticle formulation of siRNA	Liver cancer	Phase I	Alnylam Pharmaceuticals
OSI-7904 L	Liposomal thymidylate synthase inhibitor	Various cancers	Phase 11	OSI Pharmaceuticals
Combidex	Iron oxide	Tumor imaging	Phase III	Advanced Magnetics
Aurimune	Colloidal gold/TNF	Solid tumors	Phase II	Cytimmune Sciences

(continued)

Table 2 (continued)

Product/ agent	Nanoplatform	Indication	Status	Company
SGT-53	Liposome Tf anti-body/p53 gene	Solid tumors	Phase I	SynerGene Therapeutics
BIND-014	PI.GA/PLA NPs/docetaxel	Prostate cancer and others	Phase I	BIND Biosciences
AuroLase	Gold-coated silica NPs	Head and neck cancer	Phase I	Nanospectra Biosciences
Rexin-G	Targeting protein tagged phospholipid/ microRNA-122	Sarcoma, osteosarcoma, pancreatic cancer, and other solid tumors	Phase II/III (Fast Track Designation Orphan Drug Status Acquired) in the USA fully approved in the Philippines	Epeius Biotechnologies Corp.
ThermoDox	Heat-activated liposomal encapsulation of doxorubicin	Breast cancer, primary liver cancer	Approved for breast cancer; phase III for primary liver cancer	Celsion
BIND-014	Polymeric nanoparticle formulation of docetaxel	Various cancers	Phase I	BIND Bioscience
SGT53-01	Transferrin-targeted liposome with p53 gene	Solid tumors	Phase I	SynerGene Therapeutics
PEG-PGA and DON	PEG-glutaminase combined with glutamine antimetabolite 6-diazo-5-oxo-l-norleucine (DON)	Various cancers	Phase I/II	EvaluatePharma
PEG-IFN«2a	PEG-asys	Melanoma, chromic myeloid leukemia, and renal cell carcinoma, melanoma, multiple	Phase I/II	Genentech
ADI-PEG20	PEG-arginine deiminase	Hepatocellular carcinoma	Phase I	Polaris

resonance tomography as well as therapeutic carriers in cancer, cardiovascular diseases, and rheumatoid arthritis [21]. Unfortunately, the clinical trials of this particular agent were delayed due to the report of complement activation caused, presumably, by the surface present gadolinium chelates.

Lymphotropic superparamagnetic nanoparticles are being developed at the MIT-Harvard Center for Cancer Nanotechnology Excellence by Dr. Ralph Weissleder to determine if they can be used to identify undetectable lymph node metastases [22].

Drs. Robert Langer and Omid Farokhzad from MIT-Harvard CCNE developed an actively targeted nanoparticle consisting of a polymer matrix, therapeutic payload, and surface attached homing agents to facilitate accumulation in target tissue while avoiding the immune system. BIND Biosciences is the company dedicated to developing this technology. A phase I clinical study was initiated with an ascending, intravenous dose design to evaluate the safety, acceptability, and pharmacokinetics of this carrier in patients with solid tumors [23].

Dr. Sanjiv Sam Gambhir focused on the therapeutic response (Stanford University CCNE) of carbon nanotubes (CNTs) to improve colorectal cancer imaging [24].

Interestingly, the EPR effect can enable nanoparticle transport only in certain cancers tissues (e.g., inflammatory sites). Most diseased tissues are not characterized by these leaky vasculatures and, therefore, require an active mechanism of targeting to permit accumulation of nanoparticles. The reduction of uptake of nanoparticles by healthy tissues (also tissues rich with phagocytic cells) will necessitate careful designs of their size, morphology, surface characteristics, etc. The selective recognition of nanoparticles will largely depend on active ligand-enabled homing agents. Selection of a homing agent for targeting is critical and dependent on multiple variables, e.g., (1) identification of a receptor having required cell specificity, cell surface density, degree of internalization and trafficking conduit, (2) identification of an agent with ample specificity for the biological receptor, and (3) selective placement of the agent with or without a "linker" to promote the maximal projection of the ligand from the surface of the particles. Designing an "ideal" platform for imaging and therapeutics will be reliant upon careful considerations of the physicochemical characteristics of the NPs and the biology of the targeted tissue of interest. Fine-tuning will be essential to adjust the properties of these agents from initial proof-of-concept studies in vitro, ex vivo, and in vivo.

As more functionality is added to these platforms, they become multifaceted. It becomes more critical to reflect on the biocompatibility of components and the overall constructs. National Characterization Laboratory (NCL), a federally funded US government facility, is geared to assist in the biocompatibility study of these platforms for clinical translation. (Table 3). Apart from the supports from governmental agencies and laboratories, industrial entities also are slowly starting to investigate in nanotherapeutics evaluation by NCL. Major pharmaceutical companies, including AstraZeneca and Pfizer, have also invested in nanotechnology (http://cen.acs.org/articles/91/i35/Federal-Lab-Helps-Clients-Move.html) [25].

While nanotechnology offers great promise to address some of the burning biomedical issues in clinics today, the prospect of this technology in "personalized medicine" will largely be motivated by smart design principles for a translatable, "safer" platform. To understand these issues in greater details, in the following

Table 3 Nanomedicine platforms that have been evaluated by NCL

Medicine	Indication	Particle type	Company	Phase
PDS0101	Human papillomavirus-caused cancers	Positively charged liposome filled with antigen	PDS Biotechnology	Approved to begin phase I
Bind-014	Prostate cancer	Tumor-targeting polymer nanoparticle filled with docetaxel	Bind Therapeutics	Approved to begin phase II
Cyt-6091	Solid tumors	Gold nanoparticle linked to tumor necrosis factor	Cytimmune Sciences	Phase II
AuroLase	Head and neck cancer, solid tumors	Gold nanoshells with silica core	Nanospectra Biosciences	Phase I
ATI-1123	Solid tumors	Liposome filled with docetaxel	Azaya Therapeutics	Phase I complete
PNT 2258	Non-Hodgkin's lymphoma and other cancers	Liposome filled with DNA interference fragment	Pronai Therapeutics	Phase II

Sources: Companies, NCL

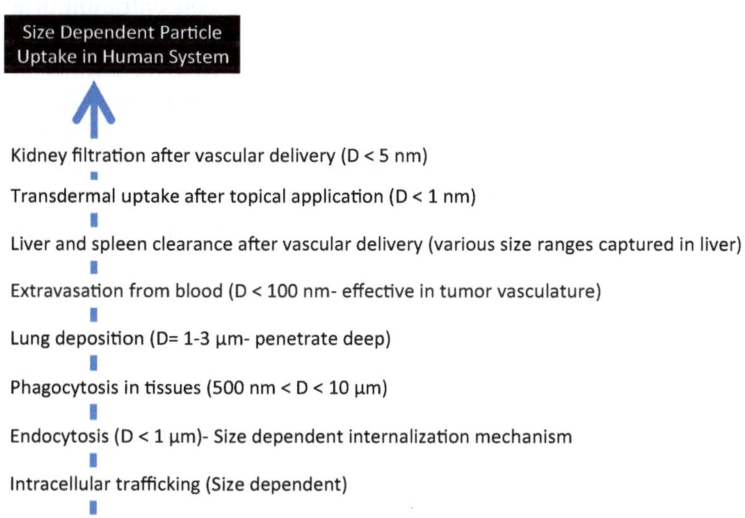

Fig. 1 Schematic representation of the size-dependent process related to particle transport in vivo (D = diameter)

sections, we will discuss the main biological barriers experienced by these nanoparticles and their physiological impacts. We realize that a better understanding of these barriers, which NPs encounter when administrated in vivo (Fig. 1), will strongly contribute to the design of more effective platforms for clinical applications in human.

2 Biological Barriers Faced by a Nanoparticle

Drug delivery faces the challenge of improving the transport of desired materials across their barriers. In a broad sense, within biological systems, natural barriers prevent foreign materials from entering the system. Small molecules with specific characteristics may overcome these barriers, but often are limited by disabling multiple barriers and do not make it to the target. Examples of these barriers include but are not limited to the BBB, nasal, skin, small intestine, and mucosal barriers. As pharmacists desire to use bigger macromolecules for drugs, the challenge of attempting to cross these barriers becomes even more prevalent. In some cases, tissue-specific transporters are utilized in carrying these larger molecules across biological barriers. However, in most of the situations, the barrier is first enfeebled such that the designed drug can permit through the targeted barrier.

Biological barriers can be classified into four broad categories: human barriers, external barriers, en route barriers, and cellular barriers (Fig. 2). The human barrier is associated with inaccuracies in measurement and how complexes behave when they encounter the first-pass mechanism. The first-pass effect is a phenomenon of the metabolism of a drug where the concentration of a drug is greatly reduced before it reaches the systemic circulation. The fraction of lost drug during the process of absorption is generally related to the liver and gut wall. This mechanism is also referred as first-pass metabolism or presystemic metabolism. The skin and mucosa make up the external barriers. En route barriers consist of blood and the extracellular matrixes, while the endosomal/lysosomal degradation and inefficient translocation to the targeted subcellular organelles comprise the cellular barrier (Fig. 3). In this introductory chapter, we will discuss these barriers in greater details.

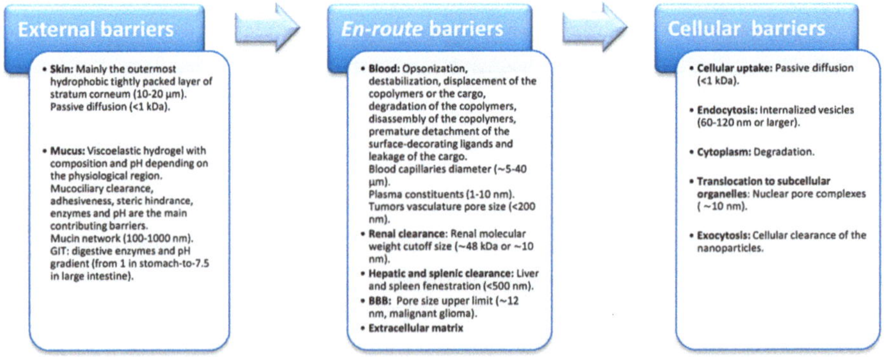

Fig. 2 Barriers towards the delivery of polymeric nanoparticles can be classified into external barriers (skin and mucosa), en route barriers (mainly destabilization and clearance in the blood and the extracellular matrixes), and cellular and subcellular barriers. Reproduced with permission from [79]

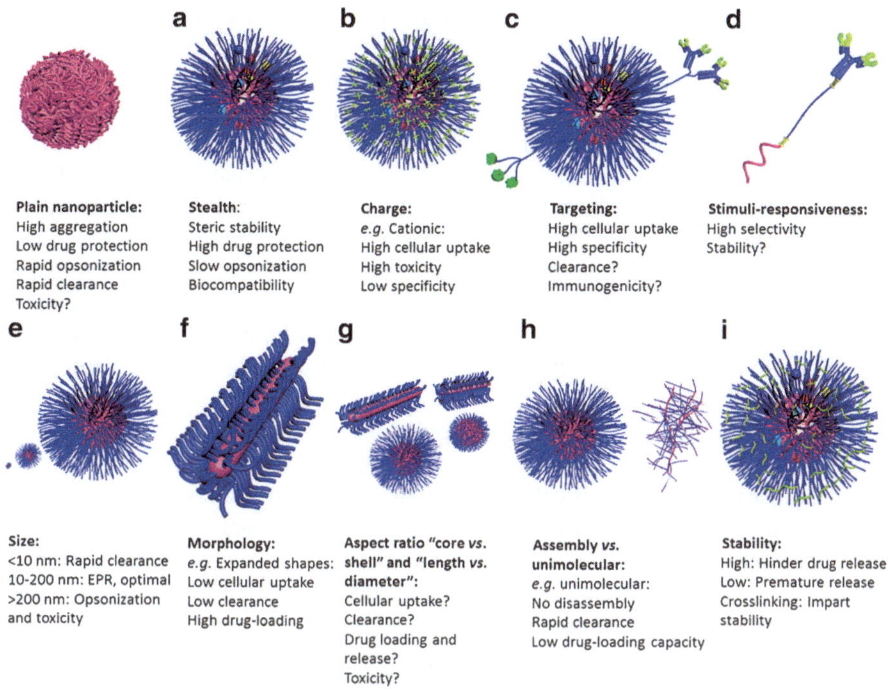

Fig. 3 Exemplified characteristics of polymeric nanoparticles: (**a**) stealth, imparts biocompatibility, steric stability, and protection of the encapsulated drug and reduces the opsonization and clearance of nanoparticles but may also reduce the cellular uptake and endosomal escape capabilities; (**b**) charge, cationic character enhances cellular uptake and endosomal escape but subject to uncontrolled tissue distribution and often associated with toxicity; (**c**) targeting, enhances cellular uptake and specificity but sometimes can accelerate the clearance and/or immunogenicity; (**d**) stimuli responsiveness, controls the dynamics of nanoparticles with possibility of releasing their cargoes at specific sites (selectivity). The stability and responsiveness of these materials under physiological and pathological conditions may vary and may result in premature release of the drug. (**e**) Size, >100 nm particles is optimal for delivery, being large enough to avoid renal clearance and small enough to reduce clearance and toxicity; (**f**) morphology, expanded morphology results in higher drug-loading capacity, lower clearance, and cellular uptake; (**g**) aspect ratio, the shell vs. core volume and length vs. diameter can greatly affect the cellular uptake, clearance, drug loading and release, and toxicity; (**h**) assembly vs. unimolecular structures, unimolecular structures are more stable (no dissociation) but can be cleared rapidly depending on the size and usually have low drug-loading capacity; and (**i**) stability, intermediate stability to circumvent physiological barriers and at the same time be able to release the drug at the target sites is required and can be achieved with different methods, for instance, by cross-linking. Reproduced with permission from [79]

2.1 External Barriers

The external barriers, skin and mucosa, affect transdermal, nasal, pulmonary, and oral administrations of nanoparticle agents. Numerous agents are delivered via oral, transdermal, pulmonary, and nasal administration routes [26] but have only shown moderate success in penetrating natural bodily defenses to achieve targeted, preferential drug, or contrast imaging agent payloads.

As the largest and most accessible organ, the skin is the first line of defense against pathogens (e.g., bacteria, viruses, toxins, parasites, and fungi) and harmful radiation exposure [27–32]. For this reason, it is inherently a challenge to deliver nanosized therapeutic and imaging agents beyond the surface to derive clinical benefit. At the same time, the skin is an attractive target for therapeutic applications due to its noninvasive administration via topical and transdermal applications, high bioavailability due to the avoidance of first-pass metabolism, and accessibility to large surface area for local therapeutic action and systemic absorption. Despite the widespread use of topical/transdermal medications, our knowledge is limited to the interaction of nanoparticles with skin conditions and the factors that determine their absorption and permeability [28–40].

The skin is comprised of three parts: epidermis, hypodermis, and the dermis. The epidermis is the most outer layer exposed to the external environment. Of the epidermis, the outermost sub-layer is called the stratum corneum which is a 10–15 μm thick stratum containing a heterogeneous mixture of keratinocytes, melanocytes, Langerhans cells, and Merkel cells. It is considered the rate-limiting barrier for diffusion which depends on the keratin content and hydration condition [40]. High keratin content and hydrated state promote permeation. Melanocytes protect us from harmful UV radiation due to their production of melanin. The function of Langerhans cells is to capture foreign entities that pose a threat to the immune system. Merkel cells are responsible for triggering our somatic senses. Keratinocytes are the most abundant cell type within the epidermis and serves as a physical barrier to prevent foreign entities from trespassing. Keratin is produced by these cells to waterproof the skin surface. Dermis, the second layer, is composed of a dense matrix of collagen, elastin, and fibrillin that forms a connective tissue layer offering mechanical support. Sweat glands, oil-secreting glands, nerve endings and hair follicles, and blood vessel networks are found in the dermis. Hypodermis, the deepest layer, is composed of adipose cells that are responsible for the accumulation of fats and regulation of body temperature. This layer is referred to the subcutaneous tissue where subcutaneous injections are administered [40].

To develop nanomedicine delivery platforms via skin requires understanding of healthy as well as diseased or injured skin conditions. Several qualitative studies have shown that nanoparticles do not penetrate healthy skin [41–49] and that skin damage tends to facilitate skin permeation of nanoparticles. However, these findings are based on a wide array of in vivo and ex vivo skin models with different nanoparticle formulations interrogated by various imaging and analytical modalities. Therefore, it is difficult to draw a consensus on the nanoparticle penetration into the skin [40]. Few studies have quantified the physicochemical properties of nanomaterials with respective to the physiological and physiopathological conditions of the skin. Based on the current findings, molecules and particles that are less than 500 Da and have a pH value of 5 will penetrate the stratum corneum [40–49]. Other factors, such as hydration condition, metabolic capacity, age, ethnicity and anatomical zone, affect the possibilities of penetration into stratum corneum. There are three penetration pathways: follicular via hair follicles and intracellular and intercellular through the lipid bilayers of the stratum corneum [11]. The outermost tightly packed layer of the stratum corneum with a thickness of 10–20 μm serves as a main external barrier as only complexes that are below 1 kDa can penetrate through it by passive diffusion.

The mucus on the other hand is a viscoelastic hydrogel in which its composition and pH depend on the physiological region of the body. The mucosal surfaces consist of a single layer of epithelial cells that line the oral, gastrointestinal, pulmonary, urogenital tracts, and the eye conjunctiva. Epithelial cells secrete mucus, a thick vicious substance organized in cross-linked, bundled, and entangled fibers. The mucus serves as a natural barrier to prevent the entry of foreign substances by entrapping it [40]. It has a higher permeability for water than skin. The structure of the mucosa is governed by mechanical stress and the anatomical zones which govern the permeability of nano-vehicles. Irritants or stimuli will trigger the mucosa to increase the amount of secretion of mucins. Mucins are glycoproteins that form the "sticky" gel-like matrix [40–49].

There are two common delivery methods to bypass the mucosal barrier by developing nanoparticles to adhere to the mucosal layer using interaction forces such as hydrogen bonding, van der Waals forces, or electrostatic interactions. Additionally, the size of the particle and molecule, concentration gradient, and non-covalent interactions determine the diffusion rate through this pathway. Second option is to engineer nanoparticles to traverse the mucosal layer by transepithelial transport or paracellular pathway via the junction complex. Hence, the nanoparticle size and surface properties are key parameters in optimizing nanocarriers to surmount the mucosal barrier.

The mucosal barrier influences pulmonary, oral, and nasal administrations. The gastric mucus composition and pH depend on the physiological region of the body. The gastric mucus, for example, consists of 90–95% water, 5–10% mucin, and about 1% electrolyte. Enzymes, nucleic acid, lipid, plasma protein, secretory IgA, bacteria, and their decomposition products make up the rest of the gastric mucus. In mucin, the amino- and carboxyl-terminal regions are rich in cysteine residues and only lightly glycosylated. It is the cysteine residues that allow for the formation of disulfide linkages among and also within mucin monomers. The problems of mucociliary clearance, adhesiveness, steric hindrance, enzymes, and pH are some of the other obstacles that are typically encountered.

2.2 *Oral Delivery*

Several different methods are available for the delivery of therapeutic drugs to the patient. Every route of administration, such as intravenous, intramuscular, and transdermal, contains its own advantage, onset, and patient comfort. Among these methods of administration, oral delivery is by far the preferred and most widely adopted mean of delivering drugs in terms of convenience and patient compliance. (http://www.cancer.gov/cancertopics/pdq/supportivecare/pain/HealthProfessional/page3#_144_toc; http://www.ncbi.nlm.nih.gov/pmc/articles/PMC3839808/).

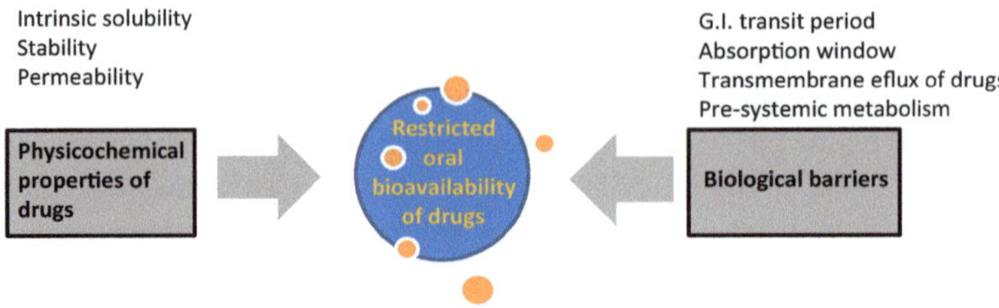

Fig. 4 A graphical representation of the challenges in the oral bioavailability of drugs

Despite the ease of administering oral delivery, there are drawbacks. There are many biological barriers that lead to a poor bioavailability. The gastrointestinal (GI) tract selectively allows the uptake of nutrients while simultaneously disallowing the uptake of pathogens and toxins. Similarly, the tract contains many protective barriers that also inhibit the uptake of drugs. For instance, the cellular and mucosal layers of the intestines can increase the instability of drugs, especially those of a peptide or protein basis. Designing a delivery system therefore has been a persistent challenge due to their several unfavorable physicochemical properties including large molecular size, susceptibility to enzymatic degradation, short plasma half-life, ion permeability, and immunogenicity. The inclination to undergo aggregation, adsorption, and denaturation is also the contributing factors that prevent successful drug delivery [50–56].

The mucosal layer of the intestines traps pathogens and particulates and rapidly clears them out – further increasing the difficulty of drug penetration. The acidic pH of the stomach and intestines degrades substances so it can additionally prevent the efficient delivery of these drugs. Luminal enzymes within the tract can also prematurely breakdown these drugs. Low permeability in the intestinal epithelial layer prevents their delivery. Additionally, the high molecular weight of macromolecular drugs inhibits their permeation via passive diffusion. Meanwhile, the solubility of the drug poses another issue. Lipid-soluble drugs can passively diffuse through the intestinal membranes, but their poor water solubility can pose problems as they try to pass the water layer in front of these membranes. A summary diagram of these challenges is seen in Fig. 4.

To enter the bloodstream, these drugs first must cross the epithelial lining of the intestines, which can be achieved through multiple pathways, including paracellular diffusion, paracellular diffusion enhanced by a modular tight junction, transcellular passive diffusion, carrier-mediated transcellular transport, transcellular diffusion modified by an apically polarized mechanism, and transcellular vesicular transport. Figure 5 depicts these different pathways. To pass through the lining of intestine cells via a transcellular pathway, the drugs must be small (define small) and lipid soluble to diffuse passively. If they contain charges or are large (define large), they will not be able to pass through. To use vesicular transport or carrier-mediated transport, the drugs must be bound to a ligand that mediates such a transportation mechanism. However, even if the drug passes out of the GI tract, it could be excreted

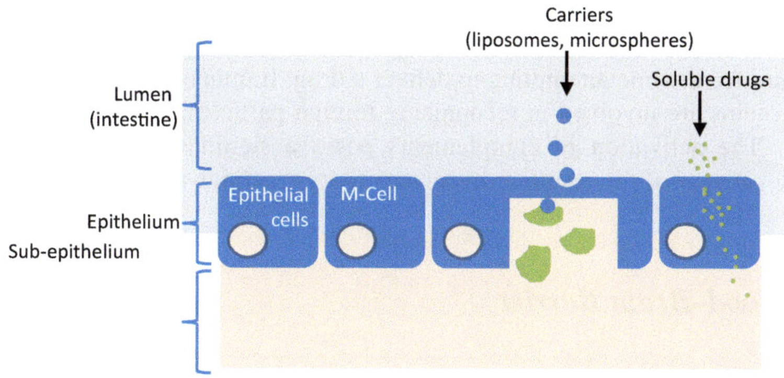

Fig. 5 Various pathways of drug absorption in the intestines. Reprinted with permission from Frontiers of Bioengineering, 2008, 38(4)

back into the intestinal lumen due to proteins like P-glycoprotein (P-gp), which is encoded by the multidrug resistance-1 (MDR-1) gene. This gene is expressed throughout the intestinal epithelia [50–56].

There are a few means to counter these barriers. This includes the incorporation P-gp inhibitors, carrier-mediated delivery, and nanoparticle encapsulation of the drugs. Various polysaccharide chains have also been found to hinder the P-gp efflux of drug compounds. A common coating for drugs includes polyethylene glycol (PEG). Although the mechanism is not yet understood, PEG was found to inhibit the secretory transport of various compounds irrespective of their molecular weight. Aside from these factors, nanoparticles used to encapsulate protein- and peptide-based drugs can also be derived from polymers like polylactic acid, polysebacic acid, and polylactic-co-glycolic acid. These nanoparticles improve the bioavailability of drugs by preventing the drug degradation from proteases, improving permeability and solubility, improving mucosal penetration, and enhancing drug targeting. While nanoparticle encapsulation can address nearly all the biological and biochemical barriers that reduce bioavailability of orally administered drugs, the production of such nanoparticles requires state-of-the-art technology as well as extensive knowledge of the physicochemistry of the drug and its delivery pathways.

2.3 *En Route Barriers*

There are various challenges encountered in transporting probes in the blood. The blood capillaries have a diameter of 5–40 μm; the tumor vasculature pore size is less than 200 nm, while the size of the plasma constituents is only between 1 and 10 nm. An immune response called opsonization can occur, which in particular refers to protein adsorption. Opsonization starts once a particle comes in contact with plasma components. Processes like opsonization, destabilization, and displacement of the cargo, degradation of the copolymer, disassembly of the copolymer, premature detachment of the surface-decorating ligand, and the leakage of payloads from cargo can occur while attempting to deliver a drug. Immunoglobulins and complement proteins are involved in recognizing foreign particles during this adsorption process. The activation of complements poses a significant challenge in drug delivery because of the resulting hypersensitivity reactions that occur [57, 58].

2.4 Blood–Brain Barrier

The blood–brain barrier (BBB) is an interface, dynamic in nature that separates the brain from the circulatory system, protecting the central nervous system from harmful substances. It also controls the transport of essential molecules and keeps the environment stable. Highly specialized endothelial cells that line brain capillaries and transduce signals from the vascular system form the BBB. The function of the BBB and its structure depend on the complex interplay between the different cell types (like the astrocytes, endothelial cells, and pericytes) and the extracellular matrix of the brain and blood flow in the capillaries. Typically, the BBB is composed of smaller subunits like biochemical dimers, transmembrane proteins, claudins, junctional adhesion molecules (JAM), and ZO-1 proteins. The BBB is one of the most difficult biological barriers to overcome in the delivery of theranostic nanoparticles. In order to cross the BBB, a disruption biochemically or by osmotic means is required. One can also provide external stimuli, e.g., localized exposure to high-intensity focused ultrasound (HIFU), to facilitate crossing the BBB. In addition, renal clearance also poses an en route barrier as complexes greater than 10–20 nm may not be excreted. Particles less than 5 nm are cleared quickly by extravasation or renal clearance.

Intercellular tight junctions between cerebral endothelial cells prevent the diffusion of hydrophilic molecules, while astrocytic perivascular end feet and basal lamina embedded with pericytes support the growth of endothelial cells. Simultaneously, they maintain structural integrity of the BBB. Small water-soluble molecules (<200 Da) can passively diffuse through the tightly packed endothelial cells comprising the BBB. Similarly, lipid-soluble molecules can also pass through it by diffusing through the membrane of the endothelial cells surrounding vessels in the brain. Mechanisms of active transport across the blood–brain barrier include adsorptive transport. Active transport occurs with charged plasma proteins that interact electrostatically with endothelial cells of the BBB and receptor-mediated transcytosis. Receptor-mediated transcytosis, on the other hand, occurs naturally for the transport of molecules such as insulin and transferrin into the brain. Furthermore, there are transport proteins that bind and actively traffic small molecules such as glucose and amino acids into the brain (Fig. 6) [59–65].

Successful delivery of theranostic nanoparticles to the brain requires designs that will utilize either active or passive transport mechanisms to cross the BBB. The injection of vasodilators widens blood vessels, which physically disrupts the

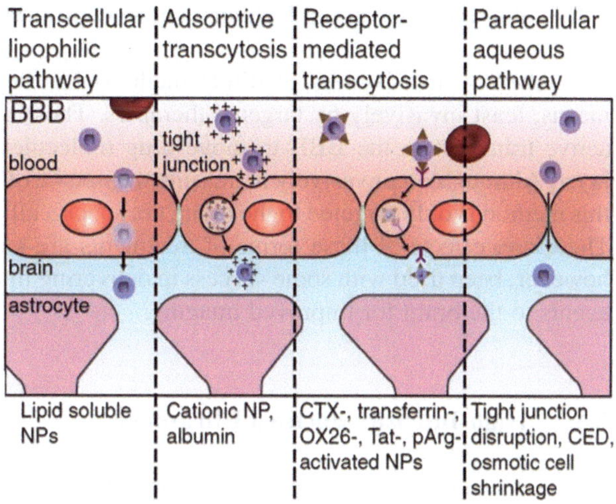

Fig. 6 Tight junctions between endothelial cells prevent passive accumulation of theranostic nanoparticles in the brain, necessitating the need for either active or disruptive mechanisms to bypass the blood–brain barrier

blood–brain barrier by increasing the size of gaps between endothelial cells and allows theranostic devices to penetrate.

In the case of magnetic nanoparticles, focused ultrasound can be used to disrupt the blood–brain barrier creating an EPR effect, while a magnetic field actively pushes the theranostic devices into the brain. However, strategies utilizing disruption of the BBB also compromise the natural defense provided by the BBB and may allow undesired foreign substances, such as bacteria and viruses, to enter the brain. Strategies utilizing knockdowns of genes involved in the formation of tight junctions in the BBB, such as claudin-5, provide transient access to the brain until expression of the knockdown gene returns. The attachment of BBB penetrating ligands to the surface of theranostic devices can be used to exploit active transport mechanisms across endothelial cells. As with genetic knockdowns, strategies utilizing active transport mechanisms do not compromise natural defense barriers.

Although the BBB may seem like an impassable barrier, there are several means of transport for nutrients, signals, and waste to travel. Controlled transfer is allowed through the tight junctions by active transport through the endothelial cell layer or by enzyme-mediated diffusion through the endothelial cells. Tight junctions function as a physical barrier, allowing larger molecules to pass by changes in cell–cell adhesion molecule expression. Active transport allows for the movement of specific molecules into the endothelial cells. Diffusion, what would seem to be the easiest way to bypass the BBB, is mediated by enzymes in the endothelial cells, which control the concentration of small molecules moving into and out of the nervous system [60–63]. The tight control of movement through the BBB has made treating brain disorders a difficult process. Researchers are developing various strategies to bypass the BBB. Some of the more common ones are described in this review.

2.4.1 Passively Targeted Delivery via Receptor Mediated Transport

The inherent invasiveness of direct methods has been avoided through several means, least invasively by targeted therapies. These methods use the endogenous active transport in the BBB to move drug molecules across the endothelial cell layer. Although noninvasively increasing the uptake of drugs to the nervous system, this method is only targeted to the brain, not specifically to the exact site, i.e., tumor. Therefore, effects of these forms of treatments are still not adequate. They have, however, been used with some success in delivering fluorophores and other contrast agents to the brain for improved imaging.

2.4.2 High-Intensity Focused Ultrasound

Another noninvasive method for delivering drug molecules across the BBB is high-intensity focused ultrasound (HIFU). This method can be used to increase the local temperature of a small area, killing any potential diseased tissue. Although relatively noninvasive with respect to surgical procedures, this form of treatment can cause broad damage to surrounding tissue. Consequentially, it is not yet approved for use by the FDA.

2.4.3 High-Intensity Focused Ultrasound Mediated Transport

Although HIFU can be a blunt tool for ablation, it can be used in conjunction with microbubbles—small molecules resonate in the ultrasound field culminating in a cavitation event. This cavitation can be focused to disrupt the endothelial tight junctions, allowing drug treatments to cross the BBB. Coupled with targeted drugs, this path may be a minimally invasive alternative to surgery or other drug treatments.

2.5 Cellular Barrier

As discussed before, drug-loaded nanoparticles ("nanocarriers") can be biologically targeted to diseased sites and be delivered following active or passive targeting approaches. Either way, they will escape from the vasculature (if systemically administered), pass through extracellular matrix (ECM), and finally reach the cells. When the carriers reach their destined cell(s), they initially encounter a significant barrier, i.e., the cell membrane. A cellular membrane is comprised of a phospholipid bilayer with embedded proteins, which selectively allows the entry of certain types of molecules into the cells. Typically, the nanocarriers are engulfed via one or more mechanisms upon touching the cell surface. The interaction of the

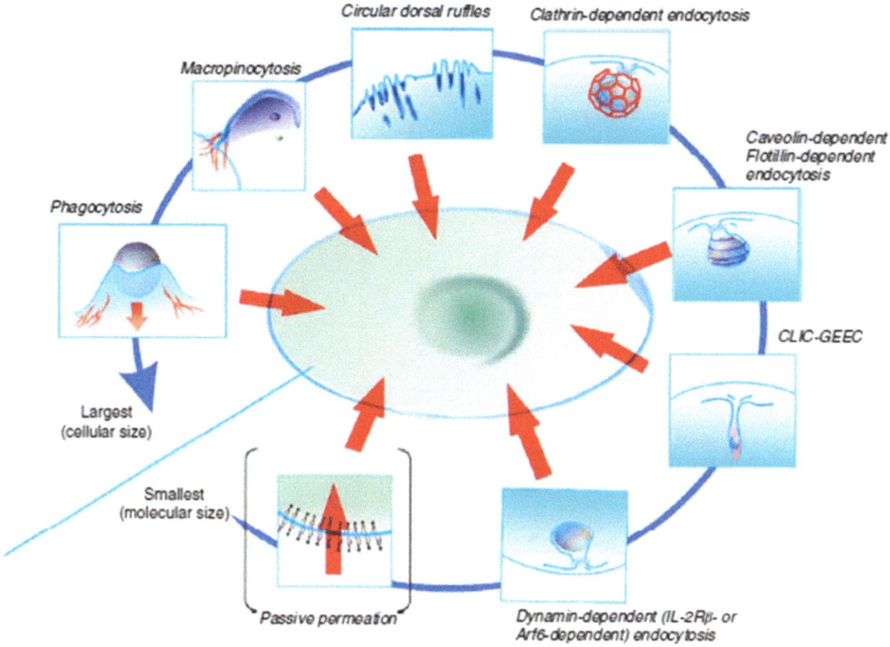

Fig. 7 Different modes of cell entry [64]

nanocarriers is highly dominated by their size, morphology, surface chemistry, and the types of cells that they are interacting with. Simple diffusion and pore transport are, however, limited for nanomedicine platforms since their cargos are relatively large (bigger than 10 nm). Thus, their internalization process is largely mediated by a mechanism known as endocytosis. The cell membrane will invaginate in order to engulf these colloidal objects with extracellular fluid and surround them into an intracellular vesicle, which develops into an endosome. Many different pathways for endocytosis are depicted in Fig. 7 [64].

The endosome is a "sorting center" for all substances internalized, which determines their fates inside the cell, ranging from their degradation and translocation into other cytoplasmic compartments to recycling. This second barrier becomes an intracellular challenge for the successful delivery of a therapeutic agent. In general, a matured endosome fuses into an acidic lysosome, and enzymatically digests engulfed foreign substances. Therefore, escaping endosomal entrapment is critical to secure the effective delivery of the drug, and, likewise, nanovectors must be engineered to interfere this processing and successfully escape this destructive event quickly.

The intracellular fates of these nanoparticles are studied both qualitatively and quantitatively. The qualitative assessment includes the visualization of fluorescently labeled nanocarriers or metallic nanoparticles under confocal microscopy or electron microscopy, respectively. The quantitative assessment can be

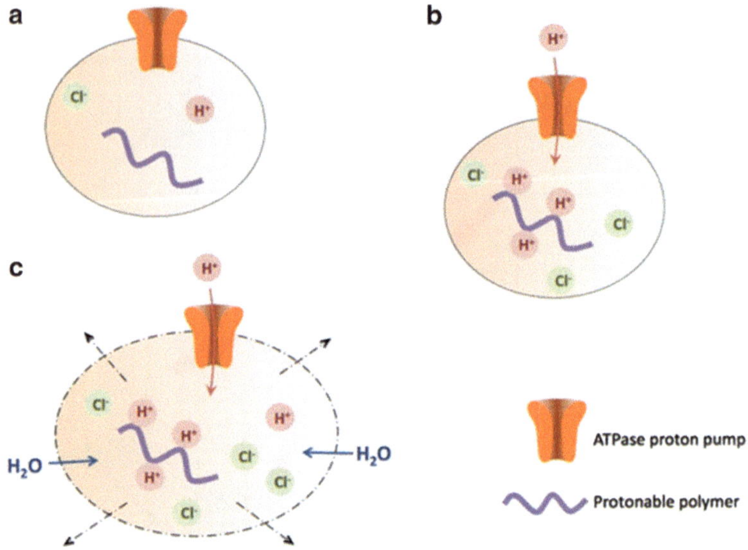

Fig. 8 Schematic of the "proton sponge" hypothesis [70]

obtained from flow cytometry employing selective endocytic pathway inhibitors [65–67].

One of the proposed endosomal/lysosomal escape mechanisms is called the "proton sponge" hypothesis (pH-buffering effect) [67–69]. Polyplex, a nanocomplex stabilized by electrostatic interaction between positively charged polymers such as polyethylenimine (PEI) or polylysine (PLL) and negatively charged molecules such as nucleic acids, is suggested to promote endosomal release when encapsulated due to its efficient pH-buffering activity. Their secondary and tertiary amino groups are protonated in the course of endosomal maturation due to the influx of hydrogen ions owing to the activity of membrane-bound ATPase protein pumps. Subsequent entry of chloride ions leads to pulling water inward, and such osmotic pressure eventually ruptures the endosome and frees the nanoparticles to cytoplasm. The steps of the proton sponge hypothesis are illustrated in Fig. 8 [67, 68].

Upon endosomal escape, the nanocarrier's payload will be unpackaged and diffused into cytoplasm. Some of the payload(s) exercise their medicinal effects in cytoplasm thus ending their journey, while others, such as genetic materials including plasmid DNA and siRNAs, continue to move to their destination, the nucleus, which houses host DNA. Therefore, the nuclear membrane is possibly the third and final intracellular barrier. The nanocargoes are believed to translocate into the nucleus when the nuclear membrane dissociates during the prophase of the cell cycle known as mitosis. Then, the translocated genetic materials will incorporate into the host DNA. If the incorporation is successful, they will start to manipulate the gene of interests by an up- or downregulation of transcriptional activity, and their therapeutic effects will take in effect [70].

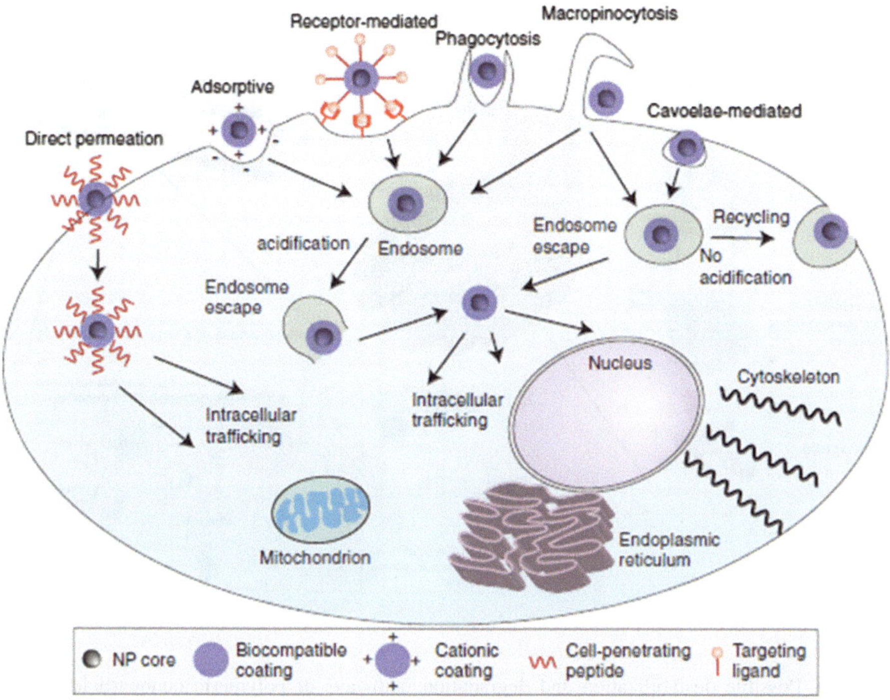

Fig. 9 Barriers encountered by theranostic nanoparticles during entry across the cell membrane

Therefore, nanocarriers will be engulfed by the cells in proximity and subsequently entrapped inside the endosome. If they successfully escape, depending on the nature of their therapeutic payloads, some might exert their effects in the cytoplasm, and others might further travel to the nucleus and control the pathways of the central dogma.

Given the importance of receptor-mediated endocytosis in bypassing the cell membrane, many targeted nanoparticle delivery strategies involve attaching a molecule to the surface of the nanoparticle that will be recognized by the surface receptors of the cell (Fig. 9). Binding of the nanoparticle to the cell surface receptor will initiate the formation of endocytic vesicles surrounding the nanoparticles that eventually bud off inside the cell. However, the size of the receptor-targeted NP may limit the effectiveness of the targeting molecule, preventing optimal uptake of NPs. NPs with a size of ~50 nm show the greatest uptake through the receptor-mediated endocytosis pathway, while NPs larger than 50 nm require longer wrapping times due to slower receptor diffusion kinetics of the cell membrane around the NP. Furthermore, subsequent intracellular trafficking of theranostic nanoparticles may be determined by the mechanism through which the nanoparticles enter the cell. Nanoparticles that enter the cell through clathrin-coated pits enter acidic endosomes where the low pH may activate enzymes that degrade the nanoparticle, while theranostic devices taken up via caveolae in lipid

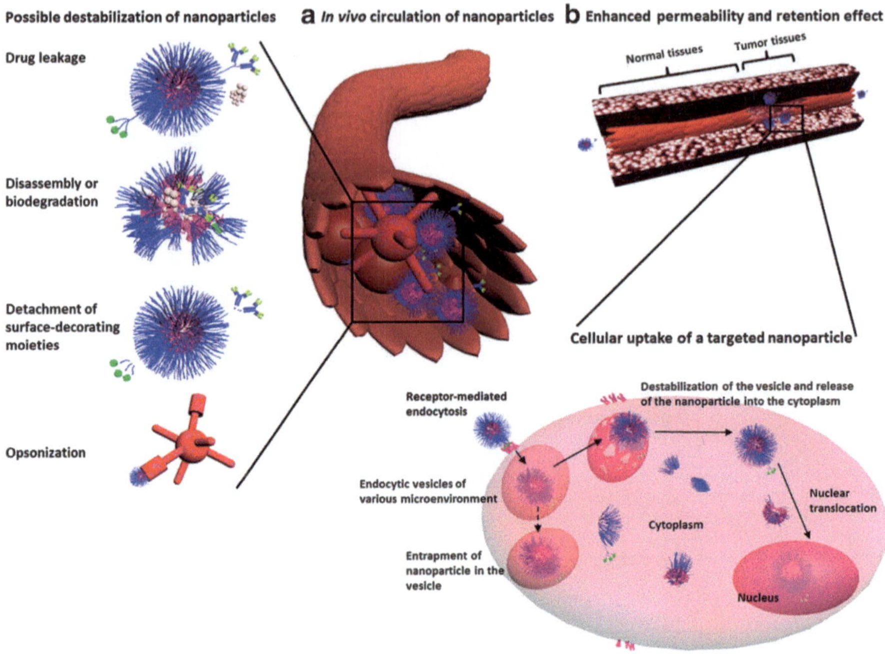

Fig. 10 Possible destabilization and degradation pathways of polymeric nanoparticles during in vivo circulation (**a**) and the EPR effect and intracellular fate of nanoparticles (**b**). Drug leakage, disassembly or degradation, detachment of surface-decorating moieties, opsonization, and clearance of nanoparticles during circulation can all be detrimental to the efficiency of nanoparticles. Tumor tissues are characterized by the leaky vasculature that allows nanoparticles to accumulate in the tumor tissues. The endocytosis of the nanoparticles can then occur via different mechanisms (e.g., via multivalent binding and receptor-mediated endocytosis), ending into endocytic vesicles of different microenvironments depending on the composition and characteristics of the nanoparticle. Entrapment of nanoparticles into the endocytic vesicles (*dashed arrow*) prevents them from reaching their target sites (cytoplasm, mitochondria, nucleus). The disassembly of polymeric nanoparticles and drug release can occur at various steps during the circulation and the intracellular trafficking pathway. Reproduced with permission from [79]

rafts may bypass transportation to endosomes or lysosomes. A complete bypass of endocytosis in favor of direct permeation through the cell membrane can be achieved through the attachment of certain cell-penetrating peptides to theranostic nanoparticles (Fig. 10) [71].

3 Conclusion

This introductory chapter is devoted broadly to the topic of theranostics and the role of functional nanometer-sized agents in personalized medicine [1, 72–74]. Over the past two decades, the field has gained a tremendous boost with the highly impending translation of bench-top sciences into clinical applications. Advancements in the areas of chemistry, molecular biology, genetics, and engineering created opportunities for interdisciplinarity with the objective of driving medical imaging and therapeutic strategies for early, sensitive detection, diagnosis, and treatment of a disease at the molecular and cellular level with uncompromised specificity [5, 7, 9, 75, 76]. A myriad of advancements has been made towards the development of defined nanostructures for performing dual functions, i.e., imaging and therapy. However, their clinical translation is still far reaching. Better understanding of biological and biophysical obstacles encountered by these agents is necessary [11, 12, 14, 77–80]. For the readers, this introductory chapter illustrates a presentation of the advancements related to this field and the biological obstacles encountered through which we hope technologies in nanomedicine applicable to translational and clinical applications will soon be booming and contribute to disentangling the elusion of disease mechanisms, therapeutic efficiency, diagnostic accuracy and safety concerns.

References

1. Shin SJ, Beech JR, Kelly KA (2012) Targeted nanoparticles in imaging: paving the way for personalized medicine in the battle against cancer. Integr Biol (Camb) 5(1):29–42
2. Pang T (2012) Theranostics, the 21st century bioeconomy and 'one health'. Expert Rev Mol Diagn 12(8):807–809
3. Lee DY, Li KC (2011) Molecular theranostics: a primer for the imaging professional. AJR Am J Roentgenol 197(2):318–324
4. James ML, Gambhir SS (2012) A molecular imaging primer: modalities, imaging agents, and applications. Physiol Rev 92(2):897–965
5. Pan D, Lanza GM, Wickline SA, Caruthers SD (2009) Nanomedicine: perspective and promises with ligand-directed molecular imaging. Eur J Radiol 70(2):274–285
6. Alberti C (2012) From molecular imaging in preclinical/clinical oncology to theranostic applications in targeted tumor therapy. Eur Rev Med Pharmacol Sci 16(14):1925–1933
7. Wang LS, Chuang MC, Ho JA (2012) Nanotheranostics–a review of recent publications. Int J Nanomed 7:4679–4695
8. Lammers T, Aime S, Hennink WE, Storm G, Kiessling F (2011) Theranostic nanomedicine. Acc Chem Res 44(10):1029–1038
9. Cabral H, Nishiyama N, Kataoka K (2011) Supramolecular nanodevices: from design validation to theranostic nanomedicine. Acc Chem Res 44(10):999–1008
10. Kunjachan S, Jayapaul J, Mertens ME, Storm G, Kiessling F, Lammers T (2012) Theranostic systems and strategies for monitoring nanomedicine-mediated drug targeting. Curr Pharm Biotechnol 13(4):609–622
11. Pan D, Caruthers SD, Chen J, Winter PM, SenPan A, Schmieder AH, Wickline SA, Lanza GM (2010) Nanomedicine strategies for molecular targets with MRI and optical imaging. Future Med Chem 2(3):471–490
12. Lanza GM (2012) ICAM-1 and nanomedicine: nature's doorway to the extravascular tissue realm. Arterioscler Thromb Vasc Biol 32(5):1070–1071
13. Wang R., Billone PS, Mullett WS (2013) Nanomedicine in action: an overview of cancer nanomedicine on the market and in clinical trials. J Nanomater 2013:Article ID 62968
14. Albanese A, Tang PS, Chan WC (2012) The effect of nanoparticle size, shape, and surface chemistry on biological systems. Annu Rev Biomed Eng 14:1–16

15. Elsabahy M, Wooley KL (2012) Design of polymeric nanoparticles for biomedical delivery applications. Chem Soc Rev 41(7):2545–2561
16. Zhang L, Gu FX, Chan JM, Wang AZ, Langer RS, Farokhzad OC (2008) Nanoparticles in medicine: therapeutic applications and developments. Clin Pharmacol Ther 83(5):761–769
17. Wagner V, Dullaart A, Bock AK, Zweck A (2006) The emerging nanomedicine landscape. Nat Biotechnol 24(10):1211–1217
18. Morton JG, Day ES, Halas NJ, West JL (2010) Nanoshells for photothermal cancer therapy. Methods Mol Biol 624:101–117
19. Davis ME, Zuckerman JE, Choi CH, Seligson D, Tolcher A, Alabi CA, Yen Y, Heidel JD, Ribas A (2010) Evidence of RNAi in humans from systemically administered siRNA via targeted nanoparticles. Nature 464(7291):1067–1070
20. Schluep T, Hwang J, Cheng J, Heidel JD, Bartlett DW, Hollister B, Davis ME (2006) Preclinical efficacy of the camptothecin-polymer conjugate IT-101 in multiple cancer models. Clin Cancer Res 12(5):1606–1614
21. Pan D, Lanza GM, Wickline SA, Caruthers SD (2009) Nanomedicine: perspective and promises with ligand-directed molecular imaging. Eur J Radiol 70(2):274–285
22. Shan G, Weissleder R, Hilderbrand SA (2013) Upconverting organic dye doped core-shell nano-composites for dual-modality NIR imaging and photo-thermal therapy. Theranostics 3 (4):267–274
23. Hrkach J, Von Hoff D, Ali M, Andrianova E, Auer J, Campbell T, De Witt D, Figa M, Figueiredo M, Horhota A, Low S, McDonnell K, Peeke E, Retnarajan B, Sabnis A, Schnipper E, Song JJ, Song YH, Summa J, Tompsett D, Troiano G, Van Geen Hoven T, Wright J, LoRusso P, Kantoff PW, Bander NH, Sweeney D, Farokhzad OC, Langer R, Zale S (2012) Preclinical development and clinical translation of a PSMA-targeted docetaxel nanoparticle with a differentiated pharmacological profile. Sci Transl Med 4:128ra39
24. Thakor AS, Gambhir SS (2013) Nanooncology: the future of cancer diagnosis and therapy. CA Cancer J Clin. doi:10.3322/caac.21199
25. Crist RM, Grossman JH, Patri AK, Stern ST, Dobrovolskaia MA, Adiseshaiah PP, Clogston JD, McNeil SE (2013) Common pitfalls in nanotechnology: lessons learned from NCI's Nanotechnology Characterization Laboratory. Integr Biol (Camb) 5(1):66–68
26. Moore T et al (2012) Nanoparticles to cross biological barriers. Biomater Sci (2012):85–121
27. Jătariu AN1, Popa M, Peptu CA (2010) Different particulate systems–bypass the biological barriers? J Drug Target 18(4):243–253
28. Potts RO, Guy RH (1993) The prediction of percutaneous penetration: a mechanistic model. In: Gurney R, Teubner A (eds) Dermal and transdermal drug delivery new insights and perspectives. Wissenschaftliche Verlagsgesellschaft mbH, Stuttgart, pp 153–160
29. Otberg N, Richter H, Schaefer H, Blume-Peytavi U, Sterry W, Lademann J (2004) Variations of hair follicle size and distribution in different body sites. J Invest Dermatol 122:14–19
30. Otberg N, Patzelt A, Rasulev U, Hagemeister T, Linscheid M, Sinkgraven R et al (2008) The role of hair follicles in the percutaneous absorption of caffeine. Br J Clin Pharmacol 65:488–492
31. Hueber F, Wepierre J, Schaefer H (1992) Role of transepidermal and transfollicular routes in percutaneous absorption of hydrocortisone and testosterone: in vivo study in the hairless rat. Skin Pharmacol 5:99–107
32. Illel B, Schaefer H, Wepierre J, Doucet O (1991) Follicles play an important role in percutaneous absorption. J Pharm Sci 80:424–427
33. Langbein L, Grund C, Kuhn C, Praetzel S, Kartenbeck J, Brandner JM et al (2002) Tight junctions and compositionally related junctional structures in mammalian stratified epithelia and cell cultures derived therefrom. Eur J Cell Biol 81:419–435
34. Nohynek GJ, Lademann J, Ribaud C, Roberts MS (2007) Grey goo on the skin? Nanotechnology, cosmetic and sunscreen safety. Crit Rev Toxicol 37:251–277

35. Lademann J, Meinke M, Sterry W, Patzelt A (2009) How safe are nanoparticles? Wie sicher sind Nanopartikel? 60:305–309
36. Bronaugh RL, Stewart RF, Congdon ER (1982) Methods for in vitro percutaneous absorption studies II. Animal models for human skin. Toxicol Appl Pharmacol 62:481–488
37. Barry BW (2002) Drug delivery routes in skin: a novel approach. Adv Drug Deliv Rev 54:31–40
38. Teichmann A, Jacobi U, Ossadnik M, Richter H, Koch S, Sterry W, Lademann J (2005) Differential stripping: determination of the amount of topically applied substances penetrated into the hair follicles. J Invest Dermatol 125:264–269
39. Feuchter D, Heisig M, Wittum G (2006) A geometry model for the simulation of drug diffusion through the stratum corneum. Comput Visualization Sci 9:117–130
40. Schneider M, Stracke F, Hansen S, Schaefer UF (2009) Nanoparticles and their interactions with the dermal barrier. Dermatoendocrinol 1(4):197–206
41. Lademann J, Richter H, Teichmann A, Otberg N, Blume-Peytavi U, Luengo J et al (2007) Nanoparticles—an efficient carrier for drug delivery into the hair follicles. Eur J Pharmaceut Biopharmaceut 66:159–164
42. Lademann J, Richter H, Schaefer UF, Blume-Peytavi U, Teichmann A, Otberg N, Sterry W (2006) Hair follicles—a long-term reservoir for drug delivery. Skin Pharmacol Physiol 19:232–236
43. Naegel A, Hansen S, Neumann D, Lehr CM, Schaefer UF, Wittum G, Heisig M (2008) Erratum to "In-silico model of skin penetration based on experimentally determined input parameters. Part II: Mathematical modelling of in-vitro diffusion experiments. Identification of critical input parameters. Eur J Pharm Biopharm 68:368–79. (doi:10.1016/j.ejpb.2007.05.018) Eur J Pharm Biopharm. 2008;68:846
44. Naegel A, Hansen S, Neumann D, Lehr CM, Schaefer UF, Wittum G, Heisig M (2008) In-silico model of skin penetration based on experimentally determined input parameters. Part II: mathematical modelling of in-vitro diffusion experiments. Identification of critical input parameters. Eur J Pharm Biopharm 68:368–379
45. Hansen S, Henning A, Naegel A, Heisig M, Wittum G, Neumann D et al (2008) In-silico model of skin penetration based on experimentally determined input parameters. Part I: experimental determination of partition and diffusion coefficients. Eur J Pharmaceut Biopharmaceut 68:352–367
46. Moghimi HR, Barry BW, Williams AC (1999) Stratum corneum and barrier performance: A model lamellar structural approach. In: Bronaugh RL, Maibach HI (eds) Drugs—cosmetics—mechanisms—methodology. Marcel Dekker, New York, Basel, Hong Kong, pp 515–553
47. Potts RO, Guy RH (1992) Predicting skin permeability. Pharmaceut Res 9:663–669
48. Magnusson BM, Anissimov YG, Cross SE, Roberts MS (2004) Molecular size as the main determinant of solute maximum flux across the skin. J Investig Dermatol 122:993–999
49. Bos JD, Meinardi MMHM (2000) The 500 Dalton rule for the skin penetration of chemical compounds and drugs. Exp Dermatol 9:165–169
50. Shaji J, Patole V (2008) Protein and peptide drug delivery: oral approaches. Indian J Pharm Sci 70(3):269–277
51. Rick S (2005) Oral protein and peptide drug delivery. In: Binghe W, Teruna S, Richard S (eds) Drug delivery: principles and applications. Wiley, New Jersey, p. 189
52. Adessi C, Sotto C (2002) Converting a peptide into a drug: strategies to improve stability and bioavailability. Curr Med Chem 9:963–978
53. Adessi C, Sotto C (2004) Strategies to improve stability and bioavailability of peptide drugs. Frontiers Med Chem 1:513–527
54. Sayani AP, Chien YW (1996) Systemic delivery of peptides and proteins across absorptive mucosae. Crit Rev Ther Drug Carrier Syst 13:85–184
55. Torres LM, Peppas NA (2000) Transmucosal delivery systems for calcitonin: a review. Biomaterials 21:1191–1196

56. O'Hagan DT, Illum L. Absorption of peptides and proteins from the respiratory tract and the potential for development of locally administered vaccine. Crit Rev Ther Drug Carrier Syst 7 (1):35–97
57. Gao H, He Q (2014) The interaction of nanoparticles with plasma proteins and the consequent influence on nanoparticles behavior. Expert Opin Drug Deliv 11(3):409–420
58. Moghimi SM (2014) Cancer nanomedicine and the complement system activation paradigm: Anaphylaxis and tumour growth. J Control Release. pii:S0168–3659(14)00217-X
59. Gidwani M, Singh AV Curr Pharm Biotechnol. 2014;14(14):1201-12. Nanoparticle Enabled Drug Delivery Across the Blood Brain Barrier: in vivo and in vitro Models, Opportunities and Challenges.
60. Rempe R, Cramer S, Qiao R, Galla HJ (2014) Strategies to overcome the barrier: use of nanoparticles as carriers and modulators of barrier properties. Cell Tissue Res 355(3):717–726
61. Alyautdin R, Khalin I, Nafeeza MI, Haron MH, Kuznetsov D (2014) Nanoscale drug delivery systems and the blood–brain barrier. Int J Nanomed 9:795–811
62. Zhang X, Zhao L, Wu J, Dong H, Xu F, Gong G, Hu Y (2012) Current advances in vehicles for brain gene delivery. Curr Gene Ther 12(5):423–436
63. Pinzón-Daza ML, Campia I, Kopecka J, Garzón R, Ghigo D, Riganti C (2013) Nanoparticle- and liposome-carried drugs: new strategies for active targeting and drug delivery across blood–brain barrier. Curr Drug Metab 14(6):625–640
64. Zaki NM, Tirelli N (2010) Gateways for the intracellular access of nanocarriers: a review of receptor-mediated endocytosis mechanisms and of strategies in receptor targeting. Expert Opin Drug Deliv 7(8):895–913
65. Hillaireau H, Couvreur P (2009) Nanocarriers' entry into the cell: relevance to drug delivery. Cell Mol Life Sci 66(17):2873–2896
66. Harush-Frenkel O, Altschuler Y, Benita S (2008) Nanoparticle-cell interactions: drug delivery implications. Crit Rev Ther Drug Carrier Syst 25(6):485–544
67. Shete HK, Prabhu RH, Patravale VB (2014) Endosomal escape: a bottleneck in intracellular delivery. J Nanosci Nanotechnol 14(1):460–474
68. Günther M, Lipka J, Malek A, Gutsch D, Kreyling W, Aigner A (2011) Polyethylenimines for RNAi-mediated gene targeting in vivo and siRNA delivery to the lung. Eur J Pharm Biopharm 77(3):438–449
69. Midoux P, Pichon C, Yaouanc JJ, Jaffrès PA (2009) Chemical vectors for gene delivery: a current review on polymers, peptides and lipids containing histidine or imidazole as nucleic acids carriers. Br J Pharmacol 157(2):166–178
70. Lam JK, Liang W, Chan HK (2012) Pulmonary delivery of therapeutic siRNA. Adv Drug Deliv Rev 64(1):1–15
71. Farkhani SM, Valizadeh A, Karami H, Mohammadi S, Sohrabi N, Badrzadeh F (2014) Cell penetrating peptides: efficient vectors for delivery of nanoparticles, nanocarriers, therapeutic and diagnostic molecules. Peptides 57C:78–94
72. Pang T (2012) Theranostics, the 21st century bioeconomy and 'one health'. Expert Rev Mol Diagn 12(8):807–809
73. Lee DY, Li KC (2011) Molecular theranostics: a primer for the imaging professional. Am J Roentgenol 197(2):318–324
74. James ML, Gambhir S (2012) S.A molecular imaging primer: modalities, imaging agents, and applications. Physiol Rev 92(2):897–965
75. Alberti C (2012) From molecular imaging in preclinical/clinical oncology to theranostic applications in targeted tumor therapy. Eur Rev Med Pharmacol Sci 16(14):1925–1933
76. Lammers T, Aime S, Hennink WE, Storm G, Kiessling F (2011) Theranostic nanomedicine. Acc Chem Res 44(10):1029–1038
77. Kunjachan S, Jayapaul J, Mertens ME, Storm G, Kiessling F, Lammers T (2012) Theranostic systems and strategies for monitoring nanomedicine-mediated drug targeting. Curr Pharm Biotechnol 13(4):609–622

78. Pan D, Pramanik M, Senpan A, Ghosh S, Wickline SA, Wang LV, Lanza GM (2010) Near infrared photoacoustic detection of sentinel lymph nodes with gold nanobeacons. Biomaterials 31(14):4088–4093
79. Elsabahy M, Wooley KL (2012) Design of polymeric nanoparticles for biomedical delivery applications. Chem Soc Rev 41(7):2545–2561
80. Sattler KD (2010) Handbook of nanophysics. CRC

Top Med Chem (2016) 20: 29–54
DOI: 10.1007/7355_2013_36

Published online: 1 March 2015

Nanoscopic Agents in a Physiological Environment: The Importance of Understanding Their Characteristics

Victoria Sherwood, Desirè Di Silvio, and Francesca Baldelli Bombelli

Abstract The application of nanotechnology in medicine signifies one of the most exciting developments in science over the last decade. Even though advancement has been made in nanoparticle engineering in terms of size, shape and surface functionalisation, the behaviour in vivo remains poorly characterised and understood. The potential impact of engineered nanomaterials on human health is strictly related to their behaviour in the biological environment. When in contact with biological fluids, nanoparticles spontaneously interact and adsorb proteins to dramatically change their surface properties. Thus, the nanoparticle surface acquires a new biological identity that will influence its stability and interaction with the cellular machinery, thereby affecting the nanoparticle biodistribution in vivo. This protein coating 'expressed' at the nanoparticle surface is what is 'read' by the cells. Consequently, methods to effectively study the structure and composition of this bio-nano interface have been emerging as key objectives in nanoscience. In this chapter, we discuss the state-of-the-art techniques for the physico-chemical characterisation of nanoparticle-protein complexes in the biological environment with particular emphasis on their impact on the efficiency and safety of a new generation of nanomedicines. We also highlight the barriers faced by nanomedicines for effective targeting and delivery in vivo.

Keywords Engineered nanomaterials, Nanomedicine, Nanoparticles, Nanotoxicology, Protein corona

---

V. Sherwood and D. Di Silvio
School of Pharmacy, University of East Anglia, Norwich Research Park, Norwich, UK

F. Baldelli Bombelli (✉)
School of Pharmacy, University of East Anglia, Norwich Research Park, Norwich, UK

CEN-European Centre for Nanomedicine c/o Dipartimento di Chimica, Materiali ed Ingegneria Chimica "Giulio Natta", Politecnico di Milano, Milan, Italy
e-mail: f.baldelli-bombelli@uea.ac.uk

Contents

Abbreviations

AFM	Atomic force microscopy
AUT	Aminoundecanoic thiol
BAM	*N-tert*-butylacrylamide
BLM	Bilayer lipid membrane
CD	Circular dichroism
DCS	Differential centrifugal sedimentation
DLS	Dynamic light scattering
ECM	Extracellular matrix
ENM	Engineered nanomaterial
EPR	Enhanced permeability and retention
FCS	Fluorescence correlation spectroscopy
FTIR	Fourier transform infrared spectroscopy
GBM	Glioblastoma
HPLC	High-performance liquid chromatography
HSA	Human serum albumin
i.v.	Intravenous
ICAM-1	Intercellular adhesion molecule-1
ICP-MS/AES	Inductively coupled plasma mass/atomic emission spectroscopy
IFP	Interstitial fluid pressure
k_{off}	Dissociation rate constant
MPS	Mononuclear phagocyte system
MS	Mass spectrometry
NIPAM	N-Isopropylacrylamide
NLS	Nuclear localisation signal
NMR	Nuclear magnetic resonance
NP	Nanoparticle(s)
PC	Protein corona
PEG	Polyethylene glycol
PS	Polystyrene
QCM	Quartz-crystal microbalance

RES	Reticuloendothelial system
SANS	Small angle neutron scattering
SAXS	Small angle X-ray scattering
SDS-PAGE	Sodium dodecyl sulphate-polyacrylamide gel electrophoresis
SEC	Size exclusion chromatography
SPR	Surface plasmon resonance
SWCNT	Single-walled carbon nanotubes
TEM	Transmission electron microscopy
Tf	Transferrin
TfR	Transferrin receptor

1 Introduction

With the advancement of nanotechnology for a widening spectrum of potential applications comes an ever-increasing production in the number of novel nanomaterials. This has resulted in the increase of nanoparticle (NP) manufacturing in recent years, and by 2020 nanotechnology is forecast to result in the annual production of around 60,000 tons of NPs [1]. Nanotechnology has shown great promise in the area of biomedical science with the development of novel multifunctional nanomaterials possess simultaneous contrast enhancement and drug carrying properties [2]. However, given the application of nanoscaled materials in many different areas such as cosmetics, the food industry, and in high-tech, it becomes crucial to study their interaction with biological substances. The contact of nanomaterials with humans can be intentional (i.e. nanomedicinal application) or unintentional (i.e. environmental exposure), where the latter has been a source of many concerns for unpredictable toxicity. The use of nanomedicines in humans requires further level of confirmation in terms of their safety. However, things are not straightforward due to the lack of proper regulatory safety guidelines for determining the hazard potential of nanoparticles [3]. The increase in number of publications in nanomedicine as well as in nanotoxicology has been exponential in the past decade. However, it is often difficult to attain conclusive results on the toxicity of engineered nanomaterials (ENMs) due to the lack complete understanding of physico-chemical properties [4].

In order to understand the potential impact of ENMs on human health, it is critical to fully characterise them in relevant biological medium [5]. When in contact with biological fluids (serum, plasma, lung lining fluids, etc.), NPs spontaneously interact and adsorb proteins to form what is known as the protein corona (PC), which dramatically changes the surface properties of NPs [6–9]. The NP's surface in a biological milieu is quickly modified by selective adsorption of proteins with the formation of a long-lived biomolecular PC constituting the primary functional interface being processed by the cellular machinery in vivo [5, 10]. Interestingly, the PC at the "bio-nano" interface is considerably different from that generated on flat surfaces of the same bulk material under the same experimental conditions [11]. There are various factors that may affect PC formation, such as the physico-chemical properties

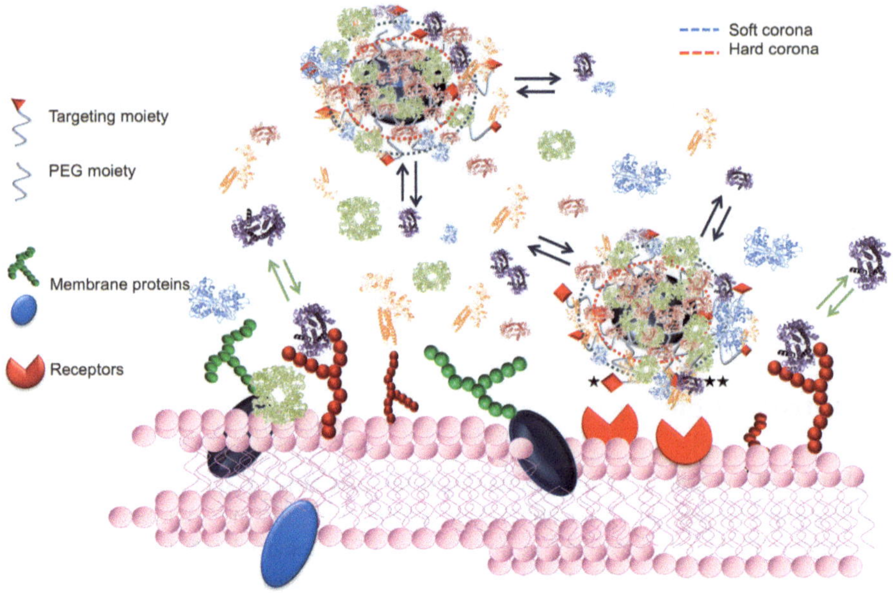

Fig. 1 Schematic drawing of the PC-NPs interacting with the cell membrane; the *red* and the *blue circles* around the NP delimit the hard corona and soft corona protein layers, respectively. The *blue arrows* indicate the dynamic nature of the PC where the adsorbed proteins exchange with the free ones. On the right, the *stars* indicate the NP surface targeting groups interacting with the receptor: (*single star*) the PC does not hinder the targeting group; (*double star*) PC hampers the binding to the receptor

of the NP, the composition of the biological fluid and the time of exposure [12]. The generally accepted view is that the PC is formed by an inner layer of more strongly bound proteins and an outer layer of proteins with less affinity to the NP surface, called the hard corona and soft corona, respectively [10]. It has been shown that the hard corona once formed and isolated from the cognate biological environment is almost irreversibly bound [10, 13], while the outer protein layers can be easily replaced. However, more studies are needed to fully elucidate the soft corona behaviour. While there are several investigations describing the dynamic process of formation of the PC in blood fluids [14, 15], which involves both protein exchange and reorganisation at the "bio-nano" interface over time, the PC evolution upon the NP journey within the body is not elucidated yet. Given that, the chemical nature, the size and shape of the NP as well as the composition of the fluid will dictate the PC characteristics; there are different simultaneous equilibria that compete for making it difficult to predict the fate of the PC in vivo (Fig. 1) [16].

The understanding of the composition, evolution and interactions exerted by PCs in vivo is critical for nanomedicines. It is evident that NPs acquire a new biological identity, which will influence their stability and interaction with living substances, thereby affecting NP bio-distribution. Moreover, their performance might be affected by the PC formation with unforeseen consequences on their efficacy. Finally, as PC is a complex protein mixture (generally between 30 and 50 different proteins), its formation involves cooperative protein-protein and protein-NP interactions with possible unfolding of one or more proteins upon adsorption with exposure of new epitopes respective to their native structures. These unravelled epitopes may interact with the cellular machinery activating cellular pathways with potential toxicity effects [17]. The latter consequence is very important for NPs that are not designed for biomedical applications as they are generally characterised by a more enhanced protein adsorption and physical instability.

In the following section, we will summarise the methodologies used to characterise the structure, composition and extrinsic function of PC-NPs highlighting the areas that need further elucidation. The biological impact of PC-NPs will be discussed with particular emphasis for drug delivery and biomedical applications. At the end, we will discuss the typical journey of intravenously injected NPs within the body experiencing the biological barriers.

2 Physico-chemical Characterisation of Nanomaterials in a Biological Environment

When NPs come in close proximity to the biological medium, they are immediately coated by a corona of biomolecules, primarily composed of proteins. The formation of this corona is a very complex process, which involves several mechanisms that ultimately cause thermodynamically favourable changes in the total enthalpy and/or entropy of the system [9]. These mechanisms can include formation of new bonds between NP and proteins, modification of the native structure of the proteins, desorption of water molecules and counter ions, and rearrangement at the interface. Finally, protein adsorption reduces the high surface energy of the bare NPs with the formation of PC-NPs whose general structure is shown in Fig. 1. The typical characteristics of PC-NPs will depend on various factors such as the physico-chemical properties of the NPs and the nature and concentration of the protein mixture. There are many experimental and theoretical studies to predict the adsorption of single proteins on nanoscaled surfaces as a function of NP size, surface functionalisation and environmental conditions [18–21]. However, mathematical modelling to predict the formation of the PC from a biological environment is extremely challenging [5].

Recently, Dell'Orco et al. [22] developed a simple mathematical model to predict the evolution of the PC formation on the basis of interaction of co-polymer-based NPs with plasma. This simple dynamical model envisages the instantaneous formation of PC-NPs with the most abundant proteins that are replaced over time by proteins with higher affinity to the NP surface and characterised by slower dissociation rate constants (k_{off}). The model can reproduce well the exchange between HSA (most abundant protein in plasma) and apo-lipoproteins (less abundant but at higher affinity) at the "nano-bio" interface of co-polymeric NPs in plasma over time observed in their experimental data. This model has been extended by Sahneh et al. [23] with the production of explicit analytical formulae to describe the dynamics of the corona composition at the initial and final states highlighted by Dell'Orco's model. However, no cooperative and/or inter-protein interaction phenomena are considered in these two models, which cannot be excluded to possess an important role in the formation and evolution of the PC-NPs. Moreover, PC-NPs through their journey within the body will encounter not only different protein composition compartments but also more complex biological structures at organ, tissue, cellular and sub-cellular levels with an impact on their structure, composition and dynamics. The implications of the interaction of PC-NPs with those structures will be discussed in the next sections.

Ascertained that ENMs interact with proteins in the biological environment acquiring an altered biological identity, it is of primary importance to determine the structure and composition of these PC-NPs and their evolution over time for understanding and predicting their biological response. Many studies have been done in the past years on the characterisation of PC-NPs in terms of size and composition, and the most conventional methodologies to investigate them are reported in Table 1. Most of the techniques used for determining the size of PC-NPs in aqueous dispersions (i.e. dynamic light scattering [DLS]) and anhydrous state (i.e. atomic force microscopy [AFM]) need to be used in a protein-free environment requiring isolation of the formed PC-NPs from the biological milieu. This involves isolation process usually consisting of centrifugation cycles and washing to obtain either hard corona or soft corona complexes depending on the number of cycles. There are few techniques, which allow the detection of the size of PC-NPs in situ in the biological environment such as differential centrifugal sedimentation (DCS) [10] and fluorescence correlation spectroscopy (FCS) [24]. The possibility to study the structure of PC-NPs in situ is a very powerful tool and permits determination if the isolated PC-NP complexes are representative of those formed in the biological environment. Obviously, any methodology has its advantages and limitations as highlighted in Table 1, and generally it is a good practice to characterise the PC-NPs using multiple techniques. The composition of PC-NPs in terms of proteins enriched in the corona requires the isolation of these complexes and the detachment of the PC from the NP surface by treatment with SDS and high temperature. Then, conventional proteomics methodologies such as SDS-PAGE and mass spectrometry are routinely used to determine the identity of the proteins. The methodologies to isolate and determine the structure and composition of PC-NPs are now well established, but there is still the need to improve in the separation of monomers PC-NPs from dimers, trimers and/or agglomerates as their composition and thereby their biological response might be different. Most of the characterisations are done on mixtures of these PC-NPs, and methodologies to separate and isolate them in sufficient amounts for further investigation are required. Moreover, centrifugation is often less effective and induces agglomeration for low density and small NPs; thus, new less invasive separation methodologies to isolate cognate PC-NPs are also desired. Finally, more sensitive methods to detect the associated proteins are needed for studying the PC of ENMs functionalised with hydrophilic polymers and/or bioactive ligands, which have less affinity to interact with the environment proteins forming a smaller corona (see Sect. 3).

Table 1 Conventional methodologies to study the properties of NP-protein complexes

Methodology	Advantage	Disadvantage	References
Structure			
Dynamic light scattering (DLS)[a]	Fast, accurate, non-perturbative	Monodispersed samples	[18, 29, 111]
Nanosight[b]	Particle tracking analysis, concentration estimation, polydispersed samples	Size limitations, material limitations	[112]
Zeta potential[a]	Fast and easy	Monodispersed samples, ionic strength limitations	[113]
Transmission electron microscopy (TEM)[a]	Small volume, high resolution for small size	Artifacts, sample preparation, electron dense samples	[28]
Atomic force microscopy (AFM)[a]	Anisotropic samples	Artifacts, difficult setting, scanning speed	[114]
Differential centrifugal sedimentation (DCS)[b]	High resolution for polydispersed samples	Density and size limitations	[115]
Fluorescence correlation spectroscopy (FCS)[b]	Low volume, low concentration	Fluorescence labelling	[29, 116]
Size exclusion chromatography (SEC)[b]	Sensitive, low volume	Sample dilution	[12]
Small angle X-ray/neutron scattering (SAXS/SANS)[b]	Shape, local structure	Complex data analysis	[117]
Composition			
Mass spectroscopy (MS)[a]	Protein identification	Expensive, tedious sample preparation	[14, 118]
SDS–PAGE (1D/2D)[a]	Qualitative and semi-quantitative analysis of mixture, cheap, fast, coupling with MS	Sensitivity depends on staining method	[117, 119]
ICP-MS/AES[a]	Low volume, concentration determination	Destructive, tedious sample preparation and calibration	[120]
X-ray absorption spectroscopy[a]	Sensitive, element microenvironment studies, no need of crystalline sample	Data interpretation, sample damage	[26]
HPLC[a]	Coupling with MS	Expensive equipment, possible coelution	[121]

(continued)

Table 1 (continued)

Methodology	Advantage	Disadvantage	References
Conformation			
UV–Vis[b]	Fast and flexible	Not quantitative or conclusive	[28]
NMR[a]	Sensitive, low volume, high resolution	Expensive, not immediate data analysis, protein size	[122]
FTIR-Raman Spectroscopy[a]	Non-destructive, complementary	Difficult data analysis	[123]
Circular Dichroism (CD)[b]	Dilute samples	Qualitative, difficult data analysis	[26]
Extrinsic functions			
Quartz crystal microbalance (QCMD)[b]	Fast, sensitive, label free detection	Complex data interpretation	[48]
Western blotting[a]	High sensitivity	Antibody dependent	[124]
Fluorescence microscopy[b]	Cell studies, specific labelling	Photo-bleaching, photo-toxicity	[125]
Protein microarrays[b]	Fast screening	Expensive, in vitro, not always correlation between activity and protein abundance	[51]
Surface plasmon resonance (SPR)[b]	Sensitivity	Immobilisation required, expensive	[111]

[a]After complexes isolation
[b]In situ analysis

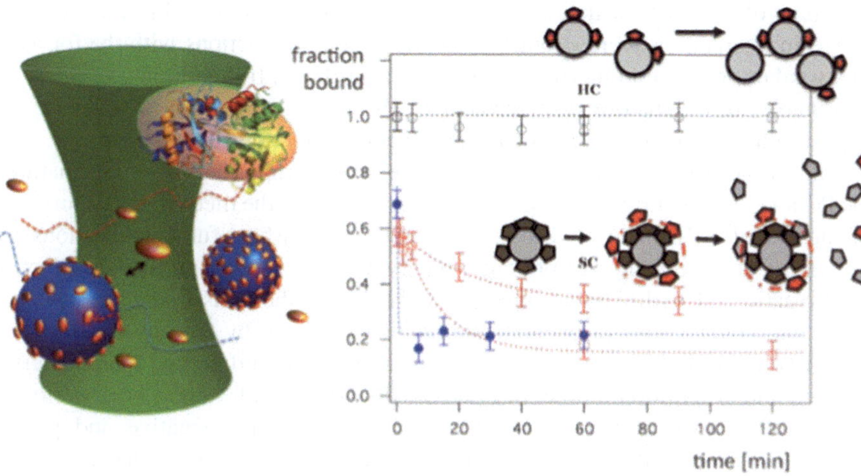

Fig. 2 Schematic representation of transferrin (Tf) PC dissociating from PS NPs measured by FCS. *Left*: Drawing of the FCS measurement volume where PC-NPs and Tf molecules diffuse in and out. *Right*: Graph reporting the bound fraction after addition of fluorescent-labelled Tf molecules to Tf PC-NPs over time—hard corona Tf-NPs did not show protein exchange while soft corona Tf-NPs showed protein exchange at the first layer level. Reprinted with permission from [13]. Copyright (2012) American Chemical Society

It has been shown for several NPs that the hard corona is strongly bound to the NP surface and proteins do not come away even at high dilution [10]. More recently the structure and dynamics of the PC has been studied by FCS [13], and NP-protein interactions have been successfully modelled for differently functionalised polystyrene (PS) NPs in the size range of 40–100 nm following a Langmuir adsorption behaviour, where NPs and proteins are the substrate and the ligand, respectively. The authors demonstrated that the PC was formed by different layers, with the first being almost irreversibly bound to the NP surface, while the outer layers could be exchanged by competitive binding of the same labelled protein or other plasma proteins (Fig. 2). However, the size and functionalisation of the NP need to be considered in the formation mechanism of the PC, and it is difficult to produce a universal model that predicts PC formation. Generally, protein adsorption on smaller NPs induces less conformational changes in the native structure of the biomolecule than on bigger surfaces, as demonstrated by the several studies carried out on the adsorption of single proteins on nanoscaled surfaces (see Table 1 and associated references). Casals et al. showed that metal and oxide NPs with sizes between 7 and 20 nm form a PC immediately in cell culture medium, which is stable for days even after dilution in water solutions [25]. Recently, Liu et al. [26] reported on the mechanism of formation of the PC on CeO_2 NPs (7 nm) showing that when the size of NPs and proteins are similar, the protein binding to the NP surface was weaker. In fact, PC CeO_2 NPs formed in serum were separated by adequate size exclusion chromatography phases (weakly negative) and showed nearly 94%

retention of the NPs with complete elution of the proteins. This indicated that NP-resin interactions were stronger than NP-protein interactions with the formation of a weaker protein corona (soft) rather than a hard one. If the binding is weak, it is reasonable to speculate a displacement of the pre-existing corona when PC-NPs approach the cell membrane, which can affect the internalisation process.

The surface functionalisation of NPs also exerts an important role on the nature of the PC and its evolution. Many studies have investigated the mechanism of formation of the PC of NPs made of different materials and surface functionalisations in a biological environment. As proteins mainly tend to adsorb on surfaces through electrostatic and hydrophobic interactions [27] with the consequent desorption of water and counter ions (a phenomenon, which is driven by entropy), cationic and more hydrophobic NPs are characterised by a stronger hard corona than negatively charged and/or hydrophilic (i.e. pegylated) NPs. Many studies have been performed on Au NPs functionalised with different ligands bearing negative and positive charges showing conflicting results. For example, it was reported that anionic Au NPs functionalised with mercaptoundecanoic acid form a transient soft PC. Size analogue cationic Au NPs functionalised with aminoundecanethiol form a strong hard corona, which keeps growing over time (rearrangement at the "bio-nano" interface) [28]. Another study showed that positive and negative polymer coated Au NPs demonstrated different biological response although qualitatively characterised by a PC of the same composition [29]. However, the structure of the PC was analysed only in terms of composition and no information about the local conformation of the proteins in the corona was assessed. Moreover, positively and negatively charged NPs exhibited different colloidal stability in the biological milieu, indicating that sometimes it is hard to decouple the diverse factors that contribute to the biological response.

In nanomedicine it is usually desired to prolong the blood circulation time of nanomaterials for increasing their chance to find the required biological target (see Sect. 4). It is known that the interaction with plasma proteins can cause opsonisation with consequent recognition of the opsonised NPs by macrophages and reduction of the blood circulation time. This was first observed for drug nanocarriers such as liposomes, where circulation times were prolonged by adding pegylated groups on the surface of the lipid vesicles, forming a coating now known as stealth corona [30–32]. This approach has been also applied to hybrid ENMs (NPs composed of both organic and inorganic components) for reducing the formation of a PC, but also for linking bioactive ligands indirectly onto the NP surface. The use of a spacer will guarantee a better arrangement of the ligands at the nano-interface [33]. This route has been widely exploited by different chemical strategies with mixed results. Pegylation of a nano-surface strongly changes the physico-chemical properties of the nanomaterial providing the NP with steric colloidal stability and a hydrophilic shell. This shell will interact with the environment proteins in different ways depending on the length and grafting density of the PEG chains. For example, it has been shown that pegylated single-walled carbon nanotubes (SWCNT) form PC complexes in plasma with different patterns as a function of the PEG conformation on the nano-surface, in particular passing from a mushroom to a mushroom-brush state

[34]. While no direct relation has been found between diverse PCs and specific biological responses passing from PEG-SWCNTs with a mushroom to a mushroom-brush conformation, the biodistribution in a murine model considerably changed. Reduced blood circulation times with higher accumulation in the spleen and a faster renal clearance were observed for the mushroom-brush conformation. Several studies have been done on pegylated Au NPs with a size range 20–80 nm, showing that smaller sizes give denser PEG coatings with a reduced uptake by the liver and the spleen in vivo, longer circulation time and better extravasation from tumour blood vessels [35, 36].

The development and optimisation of different methodologies for studying PC-NPs has allowed us to extract some general features of these protein-NP complexes.

3 Protein Corona Nanoparticles and Their Biological Function

The interaction of nanomaterials with the biological matter has been widely studied in the last 5 years as reflected from the appearance of large number of publications on this subject [5, 37]. It is now well known that nanomaterials are characterised by specific PCs when in a biological environment, but how this PC mediates the interaction with the cellular machinery still needs to be elucidated.

Generally, nanomaterials are transported into the cells through facilitated passive diffusion and/or energy-active mediated endocytosis as a function of their size and surface functional groups. There are many studies in the literature reporting on this subject and we refer the reader to these papers as a detailed report is beyond the scope of this chapter [38–40]. However, it is worth to note that recently a few groups have reported on surface-modified NPs entering the cell through alternative energy-independent mechanisms without altering irreversibly the structure of the plasma membrane [41–43]. Ultimately, these NPs seem to have cell-penetrating abilities, but the entry mechanism is not completely understood and still under study. Regarding this, it has recently been reported in an investigation on the interaction of Au NPs capped with 11-mercaptoundecane sulphonate and bilayer lipid membranes (BLM) show how the NP ability of penetrating the membrane can be tuned as a function of their size [44]. The authors developed an elegant method to quantify the NP interaction with the lipid bilayer (as model system of the cell membrane) through capacitance measurements using electrophysiology chambers separated by a hydrophobic partition with a micron-sized aperture coated with a BLM. The results were in good agreement with the biological measurements in vitro and showed an irreversibly insertion of the NPs into the bilayer, confirming the fusion of these NPs with the cell membrane. The interaction of NPs with lipid bilayers has also been explored using neutron reflectometry, which allows investigations in the local structure of the lipid membrane upon interaction with the NPs at nanometer resolution [45]. This study has shown that while cationic Au NPs tended to pass through the lipid membrane without destroying it, negative ones did not penetrate the lipid bilayer. While these methods are promising tools to study the interaction of NPs with lipid membranes, more studies are necessary to elucidate the mechanism of cell penetration of these NPs also in the presence of environment proteins whose role in the interaction with the cell surface cannot be ruled out.

A direct relation of the associated PC to the endocytosis pathway exploited by the NPs to enter the cell has not been found yet. However, it is known that the presence of a PC, independently of its specific composition, alters the affinity of the NP to the cell membrane with further consequences on the cellular uptake [46, 47]. The current understanding is that the PC decreases the surface energy of the bare NPs, reducing their association to the cell membrane. Although many studies on the interaction between the NPs and the cell membranes were performed using inhibitors and low temperature (4°C) to deactivate the endocytosis machinery as well as in serum-free conditions to exclude the PC effects, it is hard to extrapolate definitive conclusions as the biological machinery is also affected by these environmental changes. Recently, the NP-cell membrane interaction has been modelled for PS NPs of different sizes with the help of cell membrane model systems such BLMs [48]. This work combined NP-cell uptake studies with QCM measurements on BLMs in different experimental conditions. The authors found that the NP-cell membrane interaction is a two-step phenomenon, where first NPs quickly adhere to the cell membrane and successively are slowly internalised through different active mechanisms, which make the NPs enter membrane-bound endocytosis compartments. The experiments were performed on PC-NPs isolated from the excess of serum in comparison to the pristine NPs and showed that the PC decreased the non-specific interactions of the bare NPs with the cellular membrane, consequently reducing NP cellular uptake.

While the PC certainly mediates the adhesion of the NPs to the cell membrane by functioning as a coating, it is still debated if it actively promotes cellular pathways due to new and unpredictable interactions between the adsorbed proteins and cell surface receptors. We suggest that this should be evaluated case by case considering the wide range of new ENMs developed for diverse biological applications. There are few examples reporting on this aspect in the literature, though, as it is not trivial to unravel possible new active epitopes disclosed by unfolding of the adsorbed proteins enriched in the PC. An elegant study was performed on Au NPs of different sizes and functionalised with diverse polymers where the authors showed that negatively charged poly(acrylic acid)-conjugated Au NPs bound fibrinogen inducing its unfolding and interaction with the integrin receptor, Mac-1 in cells of hematopoietic lineage [49]. The fibrinogen-associated NPs were found to activate the NF-κB signalling pathway with release of inflammatory cytokines, and this biological response was tuned by NP size. Indeed, for NP sizes larger than 20 nm, the effect was massively reduced. The study has been further extended to human plasma, and NPs with different surface charges were screened. Although characterised by the same PC in terms of composition (fibrinogen was enriched in all PC-NPs), only negatively charged NPs were found to promote cytokine release from THP cells [50]. These findings demonstrate that it is important to study the extrinsic function of the PC more than its composition, as merely the presence of a protein in the PC does not necessarily mean that it will exert a biological function

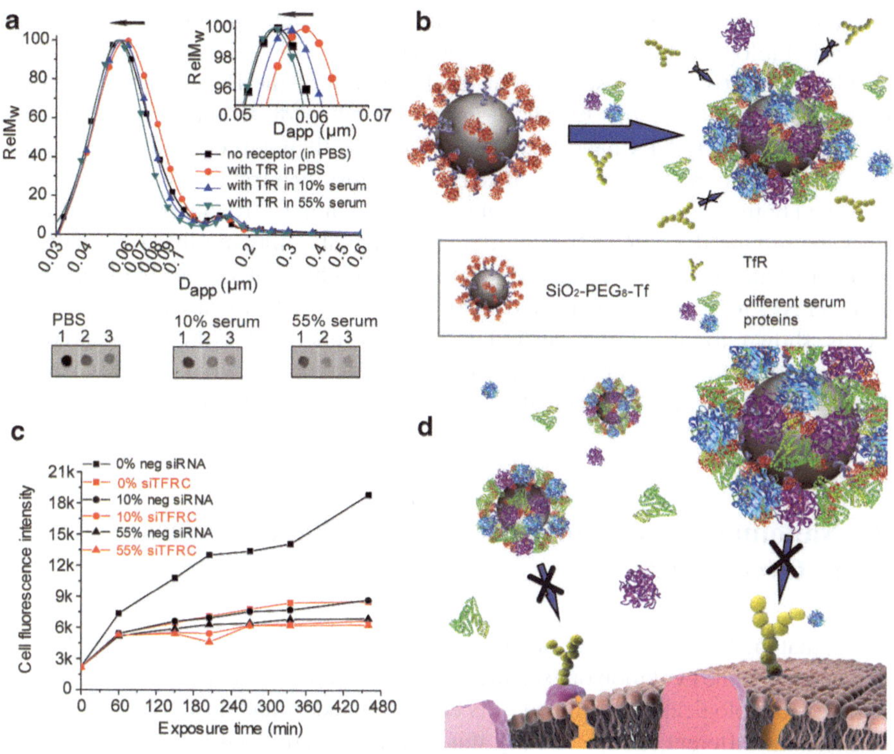

Fig. 3 Schematic drawing of SiO_2-PEG-Tf NPs interacting with the Tf receptor (TfR) in the presence of environmental proteins. (**a**) DCS graph comparing the size of SiO_2-PEG-Tf bound to the receptor in PBS and in serum where the PC hampers the binding (no change in the size). On the neat anti-Tf dot-blot are reported for Tf detection on the recovered NPs from PBS and serum. (**b**) Drawing of SiO_2-PEG-Tf NPs with the PC that hampers the binding with the TfR. (**c**) Cell fluorescence intensity due to NP uptake in serum-free and serum-containing cells in normal and TfR silenced cells. (**d**) Drawing representing the SiO_2-PEG-Tf PC NPs which cannot bind the TfR expressed on the cell membrane. With permission from [52]. Copyright 2013, Nature publishing group

on the cell. In fact, as explained above, the PC is a complex dynamic structure composed of different layers alternately bound to the NP, which likely undergoes molecular rearrangements over time and during circulation in vivo.

Protein microarrays have been used to screen possible unpredictable interactions of PC-NPs in human plasma. Carboxylated and sulphonated PS NPs functionalised with transferrin have been shown to give different binding patterns with the arrayed proteins, but no specific PC-protein interactions have been unveiled [51]. Protein microarrays represent a powerful tool to screen the extrinsic interactions displayed by PC-NPs and may be related to the unfolding of some protein upon adsorption on the NP surface with exposure of new epitopes. One limitation of this technology is the limited repertoire of proteins arrayed currently on commercially available arrays, thus reducing the variety of possible biological targets provided in vivo. The use of custom protein arrays and/or phage-display libraries for exploring the interaction displayed by PC-NPs could strongly help in unravelling possible extrinsic biological functions of these complexes.

Moreover, it has been shown that the formation of the PC can also affect the desired bioactivity of the ENMs designed for biomedical applications. In fact, pegylated Tf SiO_2 NPs have been shown to lose their targeting ability towards the Tf receptor in the presence of a PC. The loss of targeting capabilities was observed both in solution using a soluble analogue of the TfR and in vitro using cell lines expressing the TfR [52]. Figure 3 is a schematic drawing that represents the speculated mechanism through which the PC might hamper the binding of the PEG-Tf NPs to the receptor. This receptor interaction is the first evidence where the PC is shown to affect the desired biological function of the engineered NP (i.e. targeting), with clear implication for its efficiency in vivo. Further barriers faced by NPs used for biomedical applications are explained in the next section.

4 Nanomaterials' Transport In Vivo: What Barriers to Overcome?

Given that the PC provides the "identity" of the NP in a biological system [9, 10, 53], understanding the evolution or exchange of biomolecules within the PC when NPs are exposed to biological environment, it is critical to understand how NPs interact with cells within different tissues. Clearly this is a key determinant in developing tissue targeted drug nanocarriers or nanosized diagnostic tools that can be used systemically in patients' tumours. Thus, complexity of biological environments has the potential to dramatically alter PC development during targeted delivery of NPs to disseminated tumours. The focus of this section is to examine how the systemic transport of nanomedicines in patients can potentially hinder successful NP targeting in the variety of physiological environments encountered during their delivery, following intravenous (i.v.) administration of nanomedicines, using cancer patients with solid tumours as an example. As summarised in Fig. 4, i.v. administration of tumour cell-targeted NPs in patients will incur the following transport processes: blood-borne transport, transvascular transport, interstitial transport through the extracellular matrix (ECM), target cell uptake and post-delivery clearance. We will discuss each of these processes with the aim to highlight future research efforts that are required to overcome these barriers.

4.1 Blood-Borne Transport

i.v. administration of NPs may induce the formation a PC produced from the blood plasma, functionalizing the NPs in response to exposure in the blood vasculature. Within the bloodstream environment, the mononuclear phagocyte system (MPS;

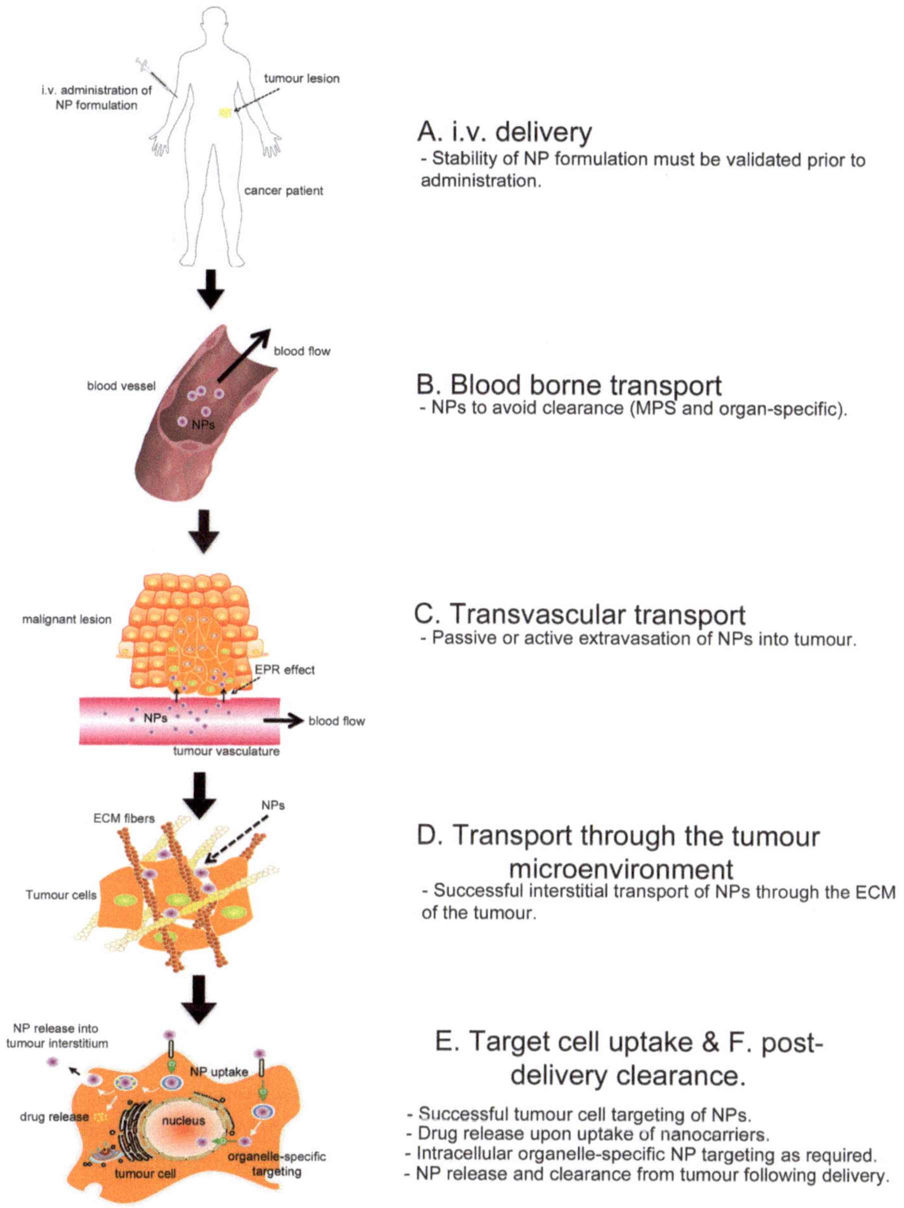

Fig. 4 Possible transport routes of nanomedicines to tumours in cancer patients. Detailed discussion of each of the transport steps is provided in the text. (**a**) Intravenous (i.v.) administration requires that stable nanoparticle (NP) formulations in the blood plasma are produced. (**b**) NPs that are not cleared by the mononuclear phagocyte system (MPS) may be delivered to the tumour. (**c**) Once at the tumour site, NPs can extravasate into the tumour tissue either by targeting the diseased vasculature or through the passive enhanced permeability and retention (EPR) effect. (**d**) Following accumulation in the tumour, NPs must then penetrate the tumour microenvironment

sometimes also referred to as the reticuloendothelial system, RES) is active. The MPS is a constituent of the immune system primarily comprising of monocytes, macrophages and dendritic cells that are responsible for phagocytosis and removal of blood-borne particulates, including i.v.-delivered NPs. Serum-induced PC formation on NPs (opsonisation) tags the particles for clearance by the MPS. Large, cationic and/or hydrophobic NP preparations are particularly susceptible and are rapidly targeted for removal by the MPS in the blood, with subsequent accumulation in the liver and spleen (see Sect. 2). It has been known as early as the 1990s that functionalizing the surface of liposomal agents with PEG molecules could dramatically enhance prolonged exposure and tumour delivery of the NPs in cancer patients [54–56]. However, pegylation also compromises NP interaction with target cells [57, 58], thus inhibiting the targeted accumulation of nanomedicines in diseased tissues. To address this issue, adaptive NPs capable of depegylation at the desired target tissue, either through external stimulation or over time, have recently been designed [59].

Organs such as the liver and spleen possess phagocytic cell populations that can remove particulates from the blood flowing into them, causing the accumulation of NPs in these organs. Alarmingly, this accumulation can often be several magnitudes higher than NP accumulation in the desired target site such as the tumour [60–62]. NP size is a critical consideration for overcoming these clearance barriers and increases plasma half-life of nanomedicines. Large NPs (>100 nm) are not only easily detected by the MPS, but can be phagocytosed by splenic macrophages and hepatic Kupffer cells [63]. Small NPs can be easily cleared by alternate mechanisms including glomerular filtration [64, 65] and hepatocyte removal [64]. This suggests that NPs of an intermediate size are favourable for increased circulating half-life, but there is currently no consensus on optimum size scale. Furthermore, NP biodistribution is also dependent on additional factors such as surface charge and composition [66]. As highlighted above, half-life is also likely to be heavily dictated by the composition of the PC.

In addition to pegylation, a number of other stealth factors have been proposed as possibilities to functionalise the surface of NPs to increase plasma half-life. Low molecular weight heparin [67] and glucose [68] have also been used as stealth materials. Another particularly promising approach is the use of a self-recognition peptide derived from CD47. CD47 is a cell surface glycoprotein expressed as a putative marker of 'self' that interacts with signal regulatory protein-α (also called CD172a) on macrophages to inhibit phagocytosis [69]. Nanobeads coated with a human CD47-derived, self-recognition peptide demonstrated enhanced circulation

---

Fig. 4 (continued) to reach the cancer cells. Barriers within this environment include a stiffened extracellular matrix (ECM) and high interstitial fluid pressure (the latter not illustrated here). (**e**) Target cell uptake can be achieved when NPs interact with the tumour cells. Here the NPs can release their drug cargo if functioning as nanocarriers or can function as contrast agents to provide imaging for the treating clinicians. Organelle-specific targeting can also be engineered into the NPs. (**f**) Post-delivery clearance of NPs following successful tumour cell targeting is required to avoid unwanted NP accumulation in patients

times, reduced clearance rates and increased drug delivery to cancerous lesions in tumour-bearing mice, highlighting the stealth properties of CD47 [70].

Following i.v. administration, NPs circulate prior to reaching the target tissue (such as the tumour). However, while in circulation, most NPs get cleared from the bloodstream and organs as described. This process may range from several minutes to hours depending on the NP composition and size [71–75]. Ideally i.v.-administered NPs should not aggregate in the blood, resist MPS detection, have prolonged circulation times and should selectively target the desired tissue.

4.2 Transvascular Transport

For access to diseased tissue, intravenously administered NPs must cross the endothelial monolayer. This monolayer functions as a semi-selective barrier that regulates the movement of molecules between the vasculature and the extravascular space within tissues. The pore size within this barrier varies between tissues and is largely dependent on the presence of inter-endothelial structures such as cell-cell junctions and fenestrae. There is strong physiological evidence that a selective size limit exists anywhere between 1 and 12 nm in healthy tissues depending on the anatomical site [76]. Tumours however often lose this selective barrier owing to their abnormal and disorganised vascular architecture [77], which affords the increased accumulation of macro- and nano-sized objects within the tumour tissue compared to healthy tissue. This process is referred to as the enhance permeability and retention (EPR) effect [78].

In addition to the leaky vasculature of tumours, lymphatic drainage within different regions of the tumour is often variable, leading to poor fluid drainage in some areas [79]. This increases the EPR effect by allowing NPs that have extravasated through the leaky blood vessels in tumours to be retained due to poor lymphatic drainage. The EPR effect is a passive targeting process allowing molecules and particulates (including nanomaterials) to accumulate in tissues with disorganised blood vessels and poor fluid drainage. However, exploitation of the EPR effect for NP delivery to tumours is limited by the size of tumours that can be successfully targeted as only large lesions of 100 mm^3 or more in volume possess it, making the passive targeting of micrometastases or poorly vascularised tumours impossible [80].

To address this limitation of the EPR effect, NPs can be formulated with ligands specific to receptors over-expressed in the tumour-associated endothelia or stroma. Suggested ligands include RGD, endothelial growth factors and antibodies targeting vascular cell adhesion molecule-1, and intercellular adhesion molecule-1 (ICAM-1), which all increase the tumour-targeting capacity of NPs [81–84]. Recent work has shown that altering the shape of ligand-bearing NPs can affect vasculature binding, where PS nanorods displaying the ICAM-1 antibody exhibit higher selectivity towards the vasculature compared to PS spheres [85]. Finally, another interesting approach taken was to further enhance the EPR effect by using tumour necrosis factor-α, which can augment tumour vasculature leakiness to increase spaces within the endothelial lining, thereby improving NP extravasation and intra-tumour accumulation [86, 87]. Taken together these studies demonstrate that transvascular transport of tumour-targeting nanomaterials can be substantially enhanced through the manipulation of the tumour vasculature itself.

4.3 Transport Through the ECM

The ECM is an inflexible scaffold formed from the deposition of secreted biological materials that promotes the higher-order organisation of cell structures within tissues. It is comprised in the large part of proteins such as laminins, fibronectins, proteases, proteoglycans, collagens and hyaluronan to name a few, which cells can physically attach to. The ECM of tumours is stiffened compared to healthy tissue due to an abundance of collagen and other ECM proteins [88] and is packed with a variety of cell types in addition to the tumour cells, including fibroblasts and cells of the immune system. Furthermore, factors causing the EPR effect also lead to enhanced interstitial fluid pressure (IFP) within tumours [89]. These factors, coupled together with abnormal metabolic processes leading to reduced pH, create a distinct set of characteristic properties within the tumour compared to healthy tissues that is commonly referred to as the tumour microenvironment.

Transport of NPs through the tumour interstitium is an important consideration for successful targeting of cancer cells. Generally, drug-resistant cancer cells that represent the most favourable tumour cell population for targeting, reside in the hypoxic compartment of the tumor, which is often poorly vascularised. The dense ECM and high IFP inhibit NP penetration to the centre of tumours and ultimately results in their accumulation at the tumour periphery, with their transport through the interstitium reliant on diffusion processes [90, 91].

Various approaches have been proposed to overcome these issues, including remodelling of tumour vasculature to lower the IFP [92], priming with chemotherapeutics prior to NP administration to reduce tumour size [93] and ECM degradation using proteases such as collagenases and matrix metalloproteinases [94–96]. Again NP design such as surface charge and in particular size are also important considerations for penetration of the dense tumour interstitium, where generally smaller NPs (ranging anywhere from 2 to 50 nm depending on NP type) are better suited to penetrate poorly vascularised lesions [83, 97–99]. Such approaches offer significant promise to circumventing the barriers imposed by the impenetrable properties of the tumour microenvironment.

4.4 Target Cell Uptake

A key step in the success of anti-cancer nanotherapies is their ability to mediate effective tumour cell uptake. While free drugs can readily enter cells by either diffusion or endocytosis, NPs due to their size are more commonly limited to endocytic mechanisms (phagocytosis, macropinocytosis or receptor-mediated endocytosis; either clathrin- or caveolin-dependent). Optimisation of this uptake process in target cells is largely dependent on the cell type, the uptake mechanism employed and the characteristics of the NP. Therefore, in the development of novel NPs for biomedical applications, detailed studies are warranted to analyse uptake mechanisms of NPs to the desired cell type.

Over-expressed cell surface markers on tumour cells can be readily exploited as targeting moieties for NPs that have been functionalised with cognate ligands for these markers. These ligands include antibodies, peptides, aptamers and other molecules such as carbohydrates. Many of the targeting approaches used are directed against highly proliferative cell populations, and as such the targeting moieties are ubiquitously expressed, for example, the transferrin receptor [100, 101]. Perhaps more sophisticated targeting involves the use of moieties that possess a more restricted expression pattern to the target tumour cells. This process is being explored as a possible mechanism to improve target cell uptake of NPs in a variety of cancer types. Indeed for some of the clinically most difficult cancers to treat, nanotechnology offers the promise of highly effective targeted drug delivery. For example, in glioblastoma (GBM) the most common and aggressive primary brain malignancy, one elegant approach has been to use an antibody-conjugated iron oxide NP that targets an epidermal growth factor variant expressed only in malignant gliomas, but not the healthy brain tissue, with effective tumour targeting in a pre-clinical model of GBM [102]. Another promising effective targeting example is the use of melanocortin-1 receptor targeting in the most lethal form of skin cancer, malignant melanoma [103]. These studies highlight that if tumour-specific cell surface markers are identified, NPs can successfully target cancer cells for drug delivery and diagnostic purposes. However, as already mentioned, the ability of the targeted NPs must be optimised using in vitro cell lines representative of the tumour tissues for testing the effects of the eventual PC. Indeed pre-screening in vitro could strongly help to engineer the NPs with the best targeting properties in the biological environment. Clearly, this will not guarantee that the targeting will work in vivo as the PC might be more complex due to the different biological compartments that the NP will encounter within the body. Lundqvist et al. have recently reported that the associated PCs of co-polymeric NPs passing from serum to the cytosol undergo rearrangements in their composition, but a sort of protein fingerprint is kept throughout [104]. Although preliminary and fairly simple, this study is the first investigation on the evolution of the PC of NPs passing through different biological environments.

Another important consideration is the effective targeting of NP cargo to the desired intracellular location, for example, the delivery of nucleic acids (such as plasmids or small-interfering RNA) to the nucleus, or uncouplers, respiratory chain inhibitors and pro-apoptotic drugs to the mitochondria. This is an emerging area of nanoengineering for biomedical applications, which has led to a number of recent technologies being developed to provide intracellular organelle-specific targeting. For example, nuclear targeting can be achieved by attaching a nuclear localisation signal (NLS) peptide to the NP, as the NLS activates the importin protein to facilitate transport across the nuclear pore [105]. For example, 60 nm DNA-polylysine complexes are able to pass through nuclear pores when attached to a NLS [106]. Mitochondria can be targeted without the need for peptide ligands, by functionalizing the surface of the NP with cationic molecules to facilitate electrostatic interactions with the mitochondrial membrane [107]. Other organelles that can also be targeted for drug delivery include lysosomes [108], peroxisomes [109], and the cytosol [110].

4.5 Post-delivery Clearance

Although a variety of NPs can be used for targeted drug delivery, upon reaching the target cell the ideal particle formulation should release its drug contents. Furthermore, the carrier itself should be biodegradable to avoid unwanted accumulation. For nondegradable NPs (e.g. hard crystalline metal nanoparticles), eventual clearance is required to ensure adverse side effects are not triggered by accumulation of NPs in the patient following delivery. The likely route of such clearance is the lymphatic system. Although as this is often perturbed in the tumour tissue, the rate of inorganic NP clearance from the tumour can be unpredictable. More work is required to understand the clearance and potential post-delivery accumulation of inorganic NPs from tumours.

5 Summary, Conclusions and Outlook

The large use of nanostructured materials in an ever-increasing number of fields makes it critical to investigate their behaviour and effects on human health in the longer term. These studies should not only be limited to biomaterials designed for biological and/or medical applications, but they should be extended to all types of nano-systems that can be exposed to humans inadvertently. It is known that nanostructured materials interact with biomolecules such as proteins, sugars and hormones dispersed in the biological environment to form dynamic adducts with different composition, size and physico-chemical properties to the cognate synthetic NPs. These PC-NPs represent what interacts with the biological matter, and we have extensively motivated the importance of studying their properties and behaviour. The general structure of these PC-NPs has been modelled as formed of less strongly bound layers of proteins (soft corona) and a more internal protein layer almost irreversibly bound to the NP surface (hard corona). Clearly, this general model is not valid for all types of NPs, and controversial findings have been discussed above. While methodologies for determining the qualitative composition of PC-NPs in terms of protein identities are reasonably advanced, there is a lack of standardised protocols to separate the mixture of different PC-NPs formed by the same NP (monomer complexes from diverse protein-NP agglomerates) in the biological milieu. This would be extremely important as the biological activity of PC-NPs of different structure and potentially composition might be very different. We also stressed the importance of developing new methods to study the extrinsic functions exerted by these PC-NPs in the perspective to unravel unpredictable interactions due to conformational changes of the proteins upon adsorption on the NP surface. In fact, it is known that proteins can strongly modify their native structure when adsorbed on a surface exposing new epitopes possibly able to activate the cellular machinery. The difficulty resides in finding possible new interactions in such a complex structure as the PC associated to an NP where a large number of proteins are mixed together to form this multi-structured layer. Here we suggested possible approaches to investigate the function of PC-NPs such as the use of protein microarrays and phage-display libraries to find new possible interactions with biological targets, which can be relevant to both medical purposes and toxicological effects. Another aspect on PC-NPs still under debate is their transient nature when encountering different biological compartments within the body. There are a few studies on the role of the PC evolution within the body, but general trends on the effect of NP surface features and size on the biodistribution in vivo have been observed. It has also been demonstrated that the functionalisation of NP surface with polymeric hydrophilic chains such as PEG reduces opsonin adsorption prolonging the blood circulation time of the NPs. However, it has also been reported that pegylation does not completely inhibit the formation of the PC with possible unexpected effects such as the loss of targeting ability of the ENM. At the end, in the last section of this chapter, we analysed in detail the 'journey' that a targeted NP to a tumour has to face within the circulation. We think that the evolution of PC-NPs should be studied in different biological environments by using cell extracts, lysates and/or ex vivo tissues. Such studies involve the use of advanced microscopy and MS technologies, but they will significantly help to design more effective targeted ENMs. Overall, we believe that a better understanding of the evolution of PC-NPs through different biological compartments, which NPs encounter when administrated in vivo, will strongly contribute to design more effective ENMs for biomedical applications.

References

1. Lewinski N, Colvin V et al (2008) Cytotoxicity of nanoparticles. Small 4(1):26–49
2. Jinhao GAO, Hongwei GU et al (2009) Multifunctional magnetic nanoparticles: design, synthesis, and biomedical applications. Acc Chem Res 42(8):1097–1107
3. Johnston H, Pojana G, Zuin S et al (2013) Engineered nanomaterial risk. Lessons learnt from completed nanotoxicology studies: potential solutions to current and future challenges. Crit Rev Toxicol 43(1):1–20
4. Nyström AM, Fadeel B (2012) Safety assessment of nanomaterials: implications for nanomedicine. J Control Release 161(2):403–408
5. Nel AE, Mädler L et al (2009) Understanding biophysicochemical interactions at the nano-bio interface. Nat Mater 8(7):543–557
6. Aggarwal P, Hall JB et al (2009) Nanoparticle interaction with plasma proteins as it relates to particle biodistribution, biocompatibility and therapeutic efficacy. Adv Drug Deliv Rev 61(6):428–437

7. Casals E, Puntes VF (2012) Inorganic nanoparticle biomolecular corona: formation, evolution and biological impact. Nanomedicine 7(12):1917–1930
8. Mahmoudi M, Lynch I et al (2011) Protein–nanoparticle interactions: opportunities and challenges. Chem Rev 111(9):5610–5637
9. Walkey CD, Chan WCW (2012) Understanding and controlling the interaction of nanomaterials with proteins in a physiological environment. Chem Soc Rev 41(7):2780–2799
10. Walczyk D, Bombelli FB et al (2010) What the cell "sees" in bionanoscience. J Am Chem Soc 132(16):5761–5768
11. Norde W, MacRitchie F et al (1986) Protein adsorption at solid–liquid interfaces: reversibility and conformation aspects. J Colloid Interface Sci 112(2):447–456
12. Lundqvist M, Stigler J et al (2008) Nanoparticle size and surface properties determine the protein corona with possible implications for biological impacts. Proc Natl Acad Sci 105(38):14265–14270
13. Milani S, Baldelli Bombelli F et al (2012) Reversible versus irreversible binding of transferrin to polystyrene nanoparticles: soft and hard corona. ACS Nano 6(3):2532–2541
14. Mortensen NP, Hurst GB et al (2013) Dynamic development of the protein corona on silica nanoparticles: composition and role in toxicity. Nanoscale 5(14):6372–6380
15. Pompa PP, Vecchio G et al (2011) Physical assessment of toxicology at nanoscale: nano dose-metrics and toxicity factor. Nanoscale 3(7):2889–2897
16. Lynch I, Dawson KA (2008) Protein-nanoparticle interactions. Nano Today 3(1–2):40–47
17. Kim ST, Saha K et al (2013) The role of surface functionality in determining nanoparticle cytotoxicity. Acc Chem Res 46(3):681–691
18. Huang R, Carney RP et al (2013) Protein-nanoparticle interactions: the effects of surface compositional and structural heterogeneity are scale dependent. Nanoscale 5:6928–6935
19. Roach P, Farrar D et al (2006) surface tailoring for controlled protein adsorption: effect of topography at the nanometer scale and chemistry. J Am Chem Soc 128(12):3939–3945
20. Shang L, Wang Y et al (2007) pH-Dependent protein conformational changes in albumin: gold nanoparticle bioconjugates: a spectroscopic study. Langmuir 23(5):2714–2721
21. Deng ZJ, Liang M et al (2012) Molecular interaction of poly(acrylic acid) gold nanoparticles with human fibrinogen. ACS Nano 6(10):8962–8969
22. Dell'Orco D, Lundqvist M et al (2010) Modeling the time evolution of the nanoparticle-protein corona in a body fluid. PLoS One 5(6):e10949
23. Darabi Sahneh F, Scoglio C et al (2013) Dynamics of nanoparticle-protein corona complex formation: analytical results from population balance equations. PLoS One 8(5):e64690
24. Röcker C, Pötzl M et al (2009) A quantitative fluorescence study of protein monolayer formation on colloidal nanoparticles. Nat Nanotechnol 4(9):577–580
25. Casals E, Pfaller T et al (2011) Hardening of the nanoparticle–protein corona in metal (Au, Ag) and oxide (Fe_3O_4, CoO, and CeO_2) nanoparticles. Small 7(24):3479–3486
26. Liu W, Rose J et al (2013) Protein corona formation for nanomaterials and proteins of a similar size: hard or soft corona? Nanoscale 5(4):1658–1668
27. Prime KL, Whitesides GM (1991) Self-assembled organic monolayers: model systems for studying adsorption of proteins at surfaces. Science 252(5010):1164–1167
28. Casals E, Pfaller T et al (2010) Time evolution of the nanoparticle protein corona. ACS Nano 4(7):3623–3632
29. Hühn D, Kantner K et al (2013) Polymer-coated nanoparticles interacting with proteins and cells: focusing on the sign of the net charge. ACS Nano. doi:10.1021/nn3059295
30. Gabizon A, Shmeeda H et al (2003) Pharmacokinetics of pegylated liposomal doxorubicin. Clin Pharmacokinet 42(5):419–436
31. Harris JM, Martin NE et al (2001) Pegylation: a novel process for modifying pharmacokinetics. Clin Pharmacokinet 40(7):539–551
32. Immordino ML, Dosio F et al (2006) Stealth liposomes: review of the basic science, rationale, and clinical applications, existing and potential. Int J Nanomed 1(3):297–315
33. Karakoti AS, Das S et al (2011) PEGylated inorganic nanoparticles. Angew Chem Int Ed 50(9):1980–1994

34. Sacchetti C, Motamedchaboki K et al (2013) Surface polyethylene glycol conformation influences the protein corona of polyethylene glycol-modified single-walled carbon nanotubes: potential implications on biological performance. ACS Nano 7(3):1974–1989
35. Shah NB, Vercellotti GM et al (2012) Blood-nanoparticle interactions and in vivo biodistribution: impact of surface peg and ligand properties. Mol Pharmaceutics 9(8):2146–2155
36. Zhang G, Yang Z et al (2009) Influence of anchoring ligands and particle size on the colloidal stability and in vivo biodistribution of polyethylene glycol-coated gold nanoparticles in tumor-xenografted mice. Biomaterials 30(10):1928–1936
37. Mailänder V, Landfester K (2009) Interaction of nanoparticles with cells. Biomacromolecules 10(9):2379–2400
38. Chithrani BD, Ghazani AA et al (2006) Determining the size and shape dependence of gold nanoparticle uptake into mammalian cells. Nano Lett 6(4):662–668
39. Sahay G, Alakhova DY et al (2010) Endocytosis of nanomedicines. J Control Release 145(3):182–195
40. Zhang S, Li J et al (2009) Size-dependent endocytosis of nanoparticles. Adv Mater 21(4):419–424
41. Dubavik A, Sezgin E et al (2012) Penetration of amphiphilic quantum dots through model and cellular plasma membranes. ACS Nano 6(3):2150–2156
42. Kostarelos K, Lacerda L et al (2007) Cellular uptake of functionalized carbon nanotubes is independent of functional group and cell type. Nat Nanotechnol 2(2):108–113
43. Verma A, Uzun O et al (2008) Surface-structure-regulated cell-membrane penetration by monolayer-protected nanoparticles. Nat Mater 7(7):588–595
44. Carney RP, Astier Y et al (2012) Electrical method to quantify nanoparticle interaction with lipid bilayers. ACS Nano 7(2):932–942
45. Tatur S, Maccarini M et al (2013) Effect of functionalized gold nanoparticles on floating lipid bilayers. Langmuir 29(22):6606–6614
46. Safi M, Courtois J et al (2011) The effects of aggregation and protein corona on the cellular internalization of iron oxide nanoparticles. Biomaterials 32(35):9353–9363
47. Ehrenberg MS, Friedman AE et al (2009) The influence of protein adsorption on nanoparticle association with cultured endothelial cells. Biomaterials 30(4):603–610
48. Lesniak A, Salvati A et al (2013) Nanoparticle adhesion to the cell membrane and its effect on nanoparticle uptake efficiency. J Am Chem Soc 135(4):1438–1444
49. Deng ZJ, Liang M et al (2011) Nanoparticle-induced unfolding of fibrinogen promotes Mac-1 receptor activation and inflammation. Nat Nanotechnol 6(1):39–44
50. Deng ZJ, Liang M, Toth I, Monteiro M, Minchin RF (2012) Plasma protein binding of positively and negatively charged polymer-coated gold nanoparticles elicits different biological responses. Nanotoxicology 7(3):314–322
51. Pitek AS, O'Connell D et al (2012) Transferrin coated nanoparticles: study of the bionano interface in human plasma. PLoS One 7(7):e40685
52. Salvati A, Pitek AS et al (2013) Transferrin-functionalized nanoparticles lose their targeting capabilities when a biomolecule corona adsorbs on the surface. Nat Nanotechnol 8(2):137–143
53. Monopoli MP, Aberg C et al (2012) Biomolecular coronas provide the biological identity of nanosized materials. Nat Nanotechnol 7(12):779–786
54. Allen TM, Hansen C (1991) Pharmacokinetics of stealth versus conventional liposomes: effect of dose. Biochim Biophys Acta 1068(2):133–141
55. Drummond DC, Meyer O et al (1999) Optimizing liposomes for delivery of chemotherapeutic agents to solid tumors. Pharmacol Rev 51(4):691–743
56. Storm G, Belliot SO et al (1995) Surface modification of nanoparticles to oppose uptake by the mononuclear phagocyte system. Adv Drug Deliv Rev 17(1):31–48
57. Hong RL, Huang CJ et al (1999) Direct comparison of liposomal doxorubicin with or without polyethylene glycol coating in C-26 tumor-bearing mice: is surface coating with polyethylene glycol beneficial? Clin Cancer Res 5(11):3645–3652

58. Song LY, Ahkong QF et al (2002) Characterization of the inhibitory effect of PEG-lipid conjugates on the intracellular delivery of plasmid and antisense DNA mediated by cationic lipid liposomes. Biochim Biophys Acta 1558(1):1–13
59. Yoo JW, Doshi N et al (2011) Adaptive micro and nanoparticles: temporal control over carrier properties to facilitate drug delivery. Adv Drug Deliv Rev 63(14–15):1247–1256
60. Campbell RB, Fukumura D et al (2002) Cationic charge determines the distribution of liposomes between the vascular and extravascular compartments of tumors. Cancer Res 62(23):6831–6836
61. Chanda N, Kattumuri V et al (2010) Bombesin functionalized gold nanoparticles show in vitro and in vivo cancer receptor specificity. Proc Natl Acad Sci U S A 107(19):8760–8765
62. Zhang XD, Wu D et al (2011) Size-dependent in vivo toxicity of PEG-coated gold nanoparticles. Int J Nanomed 6:2071–2081
63. Sadauskas E, Wallin H et al (2007) Kupffer cells are central in the removal of nanoparticles from the organism. Part Fibre Toxicol 4:10
64. Choi HS, Liu W et al (2007) Renal clearance of quantum dots. Nat Biotechnol 25(10):1165–1170
65. Kobayashi H, Brechbiel MW (2004) Dendrimer-based nanosized MRI contrast agents. Curr Pharm Biotechnol 5(6):539–549
66. Longmire M, Choyke PL et al (2008) Clearance properties of nano-sized particles and molecules as imaging agents: considerations and caveats. Nanomed Nanotechnol Biol Med 3(5):703–717
67. Socha M, Bartecki P et al (2009) Stealth nanoparticles coated with heparin as peptide or protein carriers. J Drug Target 17(8):575–585
68. Thasneem YM, Sajeesh S et al (2013) Glucosylated polymeric nanoparticles: a sweetened approach against blood compatibility paradox. Colloids Surf B Biointerfaces 108:337–344
69. Brown EJ, Frazier WA (2001) Integrin-associated protein (CD47) and its ligands. Trends Cell Biol 11(3):130–135
70. Rodriguez PL, Harada T et al (2013) Minimal “self” peptides that inhibit phagocytic clearance and enhance delivery of nanoparticles. Science 339(6122):971–975
71. Singh R, Pantarotto D et al (2006) Tissue biodistribution and blood clearance rates of intravenously administered carbon nanotube radiotracers. Proc Natl Acad Sci U S A 103(9):3357–3362
72. Hirn S, Semmler-Behnke M et al (2011) Particle size-dependent and surface charge-dependent biodistribution of gold nanoparticles after intravenous administration. Eur J Pharm Biopharm 77(3):407–416
73. Litzinger DC, Buiting AM et al (1994) Effect of liposome size on the circulation time and intraorgan distribution of amphipathic poly(ethylene glycol)-containing liposomes. Biochim Biophys Acta 1190(1):99–107
74. Manohar S, Ungureanu C et al (2011) Gold nanorods as molecular contrast agents in photoacoustic imaging: the promises and the caveats. Contrast Media Mol Imaging 6 (5):389–400
75. Lee H, Hoang B et al (2010) In vivo distribution of polymeric nanoparticles at the whole-body, tumor, and cellular levels. Pharm Res 27(11):2343–2355
76. Sarin H (2010) Physiologic upper limits of pore size of different blood capillary types and another perspective on the dual pore theory of microvascular permeability. J Angiogenesis Res. doi:10.1186/2040-2384-2-14
77. Carmeliet P, Jain RK (2011) Principles and mechanisms of vessel normalization for cancer and other angiogenic diseases. Nat Rev Drug Discov 6:417–427
78. Maeda H (2012) Macromolecular therapeutics in cancer treatment: the EPR effect and beyond. J Control Release. doi:10.1016/j.jconrel.2012.04.038
79. Padera TP, Stoll BR et al (2004) Pathology: cancer cells compress intratumor vessels. Nature 427(6976):695

80. Adiseshaiah PP, Hall JB et al (2010) Nanomaterial standards for efficacy and toxicity assessment. Nanomed Nanobiotechnol 2(1):99–112
81. Sugahara KN, Teesalu T et al (2010) Coadministration of a tumor-penetrating peptide enhances the efficacy of cancer drugs. Science 328(5981):1031–1035
82. Su S, Wang H et al (2013) iRGD-coupled responsive fluorescent nanogel for targeted drug delivery. Biomaterials 34(13):3523–3533
83. Lee H, Fonge H et al (2010) The effects of particle size and molecular targeting on the intratumoral and subcellular distribution of polymeric nanoparticles. Mol Pharm 7(4):1195–1208
84. Morral-Ruiz G, Melgar-Lesmes P et al (2013) Multifunctional polyurethane-urea nanoparticles to target and arrest inflamed vascular environment: a potential tool for cancer therapy and diagnosis. J Control Release. doi:10.1016/j.jconrel.2013.06.027
85. Kolhar P, Anselmo AC et al (2013) Using shape effects to target antibody-coated nanoparticles to lung and brain endothelium. Proc Natl Acad Sci U S A 110(26):10753–10758
86. Ten Hagen TL, Van Der Veen AH et al (2000) Low-dose tumor necrosis factor-alpha augments antitumor activity of stealth liposomal doxorubicin (DOXIL) in soft tissue sarcoma-bearing rats. Int J Cancer 87(6):829–837
87. Seynhaeve AL, Hoving S et al (2007) Tumor necrosis factor alpha mediates homogeneous distribution of liposomes in murine melanoma that contributes to a better tumor response. Cancer Res 67(19):9455–9462
88. Holback H, Yeo Y (2011) Intratumoral drug delivery with nanoparticulate carriers. Pharm Res 28(8):1819–1830
89. Goel S, Duda DG et al (2011) Normalization of the vasculature for treatment of cancer and other diseases. Physiol Rev 91(3):1071–1121
90. Boucher Y, Baxter LT et al (1990) Interstitial pressure gradients in tissue-isolated and subcutaneous tumors: implications for therapy. Cancer Res 50(15):4478–4484
91. Heldin CH, Rubin K et al (2004) High interstitial fluid pressure – an obstacle in cancer therapy. Nat Rev Cancer 4(10):806–813
92. Jain RK (2005) Normalization of tumor vasculature: an emerging concept in antiangiogenic therapy. Science 307(5706):58–62
93. Wang J, Lu Z et al (2011) Improving delivery and efficacy of nanomedicines in solid tumors: role of tumor priming. Nanomed Nanotechnol Biol Med 6(9):1605–1620
94. Eikenes L, Bruland OS et al (2004) Collagenase increases the transcapillary pressure gradient and improves the uptake and distribution of monoclonal antibodies in human osteosarcoma xenografts. Cancer Res 64(14):4768–4773
95. Goodman TT, Olive PL et al (2007) Increased nanoparticle penetration in collagenase-treated multicellular spheroids. Int J Nanomed 2(2):265–274
96. Wong C, Stylianopoulos T et al (2011) Multistage nanoparticle delivery system for deep penetration into tumor tissue. Proc Natl Acad Sci U S A 108(6):2426–2431
97. Cabral H, Matsumoto Y et al (2011) Accumulation of sub-100 nm polymeric micelles in poorly permeable tumors depends on size. Nat Nanotechnol 6(12):815–823
98. Huang K, Ma H et al (2012) Size-dependent localization and penetration of ultrasmall gold nanoparticles in cancer cells, multicellular spheroids, and tumors in vivo. ACS Nano 6(5):4483–4493
99. Huo S, Ma H et al (2013) Superior penetration and retention behavior of 50 nm gold nanoparticles in tumors. Cancer Res 73(1):319–330
100. Daniels TR, Delgado T et al (2006) The transferrin receptor part II: targeted delivery of therapeutic agents into cancer cells. Clin Immunol 121(2):159–176
101. Daniels TR, Delgado T et al (2006) The transferrin receptor part I: biology and targeting with cytotoxic antibodies for the treatment of cancer. Clin Immunol 121(2):144–158
102. Hadjipanayis CG, Machaidze R et al (2010) EGFRvIII antibody-conjugated iron oxide nanoparticles for magnetic resonance imaging-guided convection-enhanced delivery and targeted therapy of glioblastoma. Cancer Res 70(15):6303–6312

103. Lu W, Xiong C et al (2009) Targeted photothermal ablation of murine melanomas with melanocyte-stimulating hormone analog-conjugated hollow gold nanospheres. Clin Cancer Res 15(3):876–886
104. Lundqvist M, Stigler J et al (2011) The evolution of the protein corona around nanoparticles: a test study. ACS Nano 5(9):7503–7509
105. Wagstaff KM, Jans DA (2009) Importins and beyond: non-conventional nuclear transport mechanisms. Traffic 10(9):1188–1198
106. Chan CK, Senden T et al (2000) Supramolecular structure and nuclear targeting efficiency determine the enhancement of transfection by modified polylysines. Gene Therapy 7(19):1690–1697
107. Boddapati SV, D'Souza GG et al (2008) Organelle-targeted nanocarriers: specific delivery of liposomal ceramide to mitochondria enhances its cytotoxicity in vitro and in vivo. Nano Lett 8(8):2559–2563
108. Bareford LM, Swaan PW (2007) Endocytic mechanisms for targeted drug delivery. Adv Drug Del Rev 59(8):748–758
109. Terlecky SR, Koepke JI (2007) Drug delivery to peroxisomes: employing unique trafficking mechanisms to target protein therapeutics. Adv Drug Del Rev 59(8):739–747
110. Vasir JK, Labhasetwar V (2007) Biodegradable nanoparticles for cytosolic delivery of therapeutics. Adv Drug Deliv Rev 59(8):718–728
111. Canovi M, Lucchetti J et al (2012) Applications of surface plasmon resonance (SPR) for the characterization of nanoparticles developed for biomedical purposes. Sensors 12(12):16420–16432
112. Jedlovszky-Hajdú A, Bombelli FB et al (2012) Surface coatings shape the protein corona of SPIONs with relevance to their application in vivo. Langmuir 28(42):14983–14991
113. Thielbeer F, Johansson EMV et al (2013) Influence of spacer length on the cellular uptake of polymeric nanoparticles. Macromol Biosci 13(6):682–686
114. Natte K, Friedrich JF et al (2013) Impact of polymer shell on the formation and time evolution of nanoparticle–protein corona. Colloids Surf B Biointerfaces 104:213–220
115. Laurent S, Burtea C et al (2013) Significance of cell "observer" and protein source in nanobiosciences. J Colloid Interface Sci 392:431–445
116. Nienhaus GU, Maffre P, Nienhaus K (2013) Studying the protein corona on nanoparticles by FCS. Methods Enzymol 519:115–137
117. Wang J, Jensen UB et al (2011) Soft interactions at nanoparticles alter protein function and conformation in a size dependent manner. Nano Lett 11(11):4985–4991
118. Martin S, Manuela S-B et al (2013) Serum protein identification and quantification of the corona of 5, 15 and 80 nm gold nanoparticles. Nanotechnology 24(26):265103
119. Jansch M, Stumpf P et al (2012) Adsorption kinetics of plasma proteins on ultrasmall superparamagnetic iron oxide (USPIO) nanoparticles. Int J Pharm 428(1–2):125–133
120. Freese C, Uboldi C, Gibson MI et al (2012) Uptake and cytotoxicity of citrate-coated gold nanospheres: comparative studies on human endothelial and epithelial cells. Part Fibre Toxicol 9:23
121. Bimbo LM, Sarparanta M et al (2012) Cellular interactions of surface modified nanoporous silicon particles. Nanoscale 4(10):3184–3192
122. Calzolai L, Franchini F et al (2010) Protein–nanoparticle interaction: identification of the ubiquitin–gold nanoparticle interaction site. Nano Lett 10(8):3101–3105
123. Goy-López S, Juárez J et al (2012) Physicochemical characteristics of protein–NP bioconjugates: the role of particle curvature and solution conditions on human serum albumin conformation and fibrillogenesis inhibition. Langmuir 28(24):9113–9126
124. Zheng Q, Yang H et al (2013) The role and mechanisms of nanoparticles to enhance radiosensitivity in hepatocellular cell. Biomed Pharmacother. doi:10.1016/j.biopha.2013.04.003
125. Godugu C, Patel AR et al (2013) Inhalation delivery of telmisartan enhances intratumoral distribution of nanoparticles in lung cancer models. J Control Release. doi:10.1016/j.jconrel.2013.06.036

Top Med Chem (2016) 20: 55–74
DOI: 10.1007/7355_2014_76

Published online: 20 September 2014

Rational Design of Multifunctional Nanoscale Self-Assembled Soft Materials for Biomedical Delivery Application

Shrinivas Venkataraman

Abstract Soft matter-based self-assembled nanostructures are promising for therapeutic delivery. Recent advances in synthetic polymerisation chemistries and reactive orthogonal functionalisation strategies have enabled straightforward access to well-defined nanostructures with precise control over numerous physico-chemical properties. Ability to integrate multiple components such as imaging/contrast agent, targeting ligand and smart components on to a nanocarrier has opened up innumerable possibilities in biomedical delivery application. In this chapter, key principles in the design of multifunctional nanocarriers and the challenges with clinical translation will be presented.

Keywords Drug delivery, Nanoscale, Nanostructures, Self-assembly, Soft matter

Contents

S. Venkataraman (✉)
Institute of Bioengineering and Nanotechnology, 31 Biopolis Way, The Nanos,
Singapore 138669, Singapore
e-mail: shrinivas@ibn.a-star.edu.sg

Abbreviations

ADME	Adsorption, distribution, metabolism and excretion
API	Active pharmaceutical ingredient
ATRP	Atom transfer radical polymerisation
CAC	Critical aggregation concentration
CMC	Critical micellisation concentration
CRP	Controlled radical polymerisation
DLS	Dynamic light scattering
DNA	Deoxyribonucleic acid
EPR	Enhanced permeation and retention
FDA	Food and Drug Administration
LCST	Lower critical solution temperature
mPEG	Poly(ethylene glycol) methyl ether
NME	New molecular entity
NMR	Nuclear magnetic resonance
NMRP	Nitroxide-mediated radical polymerisation
OC	Organo-catalytic
PD	Pharmacodynamics
PDI	Polydispersity index
PEG	Poly(ethylene glycol)
PK	Pharmacokinetics
Ppm	Parts per million
R&D	Research and development
RAFT	Reversible addition-fragmentation chain transfer
RES	Reticuloendothelial system
RNA	Ribonucleic acid
ROP	Ring-opening polymerisation
ROS	Reactive oxygen species
SANS	Small-angle neutron scattering
SAXS	Small-angle X-ray scattering
SCFT	Self-consistent field theory
SEC	Size-exclusion chromatography
*si*RNA	Small interfering ribonucleic acid
TEM	Transmission electron microscopy

1 Introduction

The main objective of pharmaceutical research and development (R&D) process is to develop effective, economical and safe next-generation therapeutic agents that meet all the regulatory requirements and also at the same time provide sustainable return on investment for the investors [1]. Currently R&D sector of pharmaceutical industry is facing tremendous pressure from issues including losses of revenue due to key patent expirations (and hence competition from generic manufacturers) along with reduced R&D output in terms of successfully launched approved new molecular entity (NME) [2]. In order to improve upon the current cost-constrained health-care systems, there is a real and urgent need for pharmaceutical industries to adopt innovative revolutionary technologies in synergy with comprehensive understanding of biology of diseases, for the development of next-generation medicines without incurring unsustainable R&D costs.

Introduction of novel technologies and approaches such as high-throughput screening, combinatorial chemistry and computational chemistry, in the past, in conjunction with the drug discovery and development processes has tremendously benefited the pharma sector. Over the past decades, nanotechnology has emerged as a key add-on approach, holding incredible potential to revolutionise the drug discovery process via innovative formulation [3]. Efficacy of conventional drug discovery and development process could be significantly improved with the use of nanocarriers to solubilise relatively challenging drug candidates. There seems to be a lot of opportunities to expand the scope and efficacy of targets. Significant percent of the target candidates identified by combinatorial approaches are poorly water soluble and necessitate advanced formulation strategies. Traditionally in the drug discovery process, lead candidate is optimised not only for their potency but also for their favourable pharmacokinetics (PK) and pharmacodynamics (PD). Essentially a candidate demonstrating potent biological activity with poor water solubility would be discontinued from further development, whereas a less potent candidate with optimal pharmacological properties would be developed further. Such compromises in drug discovery and development may lead to development of less potent drugs. With the advent of nanocarriers, solubility and PK of the drug candidates can be improved by several orders without compromising on potency. Particularly, the use of nanoscale self-assembled soft materials as therapeutic carriers expands the applicability of drugs by addressing specific shortcomings associated with the free drugs. For example, with an appropriate nanocarrier, significantly higher doses of poorly water-soluble drugs can be effectively administered whilst reducing the adverse side effects. Several nanotechnology-based products currently in the market or in the development pipeline are focused on improving the efficacy and safety profiles of previously approved drugs. Also for biologics-based therapeutics including DNA, *si*RNA and proteins, nanotechnological approaches can be specifically designed to navigate the innate biological barriers (such as enzymatic degradation, uptake by reticuloendothelial system (RES) and poor cellular uptake) that are otherwise almost insurmountable [4].

Amongst different classes of nanomaterials, the design and synthetic flexibility of organic (macro)molecular amphiphiles, capable of forming nanostructures via self-assembly, render them well suited for biomedical applications. In this chapter we will highlight some of the key design considerations for the development of well-defined soft matter-based (polymer or lipid) nanostructures that exhibit tremendous potential in revolutionising biomedical delivery applications. First the design criteria and approaches to make these nanocarriers will be presented

(Sect. 2). Specific challenges associated with this nanoscale therapeutic delivery will also be presented (Sect. 3), followed by summary and outlook (Sect. 4).

2 Access to Nanostructures

Molecular self-assembly-based 'bottom-up' approach is by far the most versatile approach to access functional nanocarriers in the size ranging from ~10 nm to submicron sizes [5]. This can be readily achieved by simply designing the right molecular precursor with built-in segments capable of self-assembly without any or minimal external intervention. In contrast to the lithography-based 'top-down' approaches, self-assembly provides access to well-defined nanostructures under ambient conditions, in a cost- and energy-efficient manner by eliminating the need for huge upfront investment in the microfabrication equipments. Hence, self-assembly approach is very attractive for the development of nanoscale therapeutic delivery. A wide variety of therapeutics ranging from hydrophobic small-molecule drugs to complex hydrophilic biomacromolecules can be loaded into these self-assembled nanostructures. Apart from engineering the forces, governing the assembly, materials could also be designed to disassemble in a programmed fashion with the use of built-in degradable or stimuli-responsive chemistries [6, 7]. Such programmable (dis)assembly would be useful in modulating encapsulation and eventual release of the encapsulated cargoes.

Though a variety of amphiphilic precursors could be in principle used to form of self-assembled nanostructures, amphiphilic block copolymers constitute an important and versatile class of materials [8, 9]. Compared to conventional surfactants, these block copolymers offer great opportunity to tailor numerous aspects of nanostructural physico-chemical properties [10, 11]. A multifunctional nanocarrier derived from block copolymer core-shell nanostructure cannot only be used to encapsulate therapeutics but also can be integrated with other functionalities such as imaging agent, ligands for active targeting, etc. (Fig. 1) [12]. In this section, approaches to synthesise block copolymers, their self-assembly and shape control, will be discussed.

2.1 Precision Syntheses of Amphiphilic Precursors

Recent advancements in synthetic polymer chemistry along with developments in highly efficient orthogonal ligation strategies [13] have empowered chemists to design amphiphilic systems with practically innumerable possibilities. In the past few decades, several strategies to exert precise control over numerous aspects of polymer composition have emerged. A number of mechanistically distinct controlled radical polymerisation (CRP) [14] methodologies such as nitroxide-mediated radical polymerisation (NMRP) [15], atom transfer radical

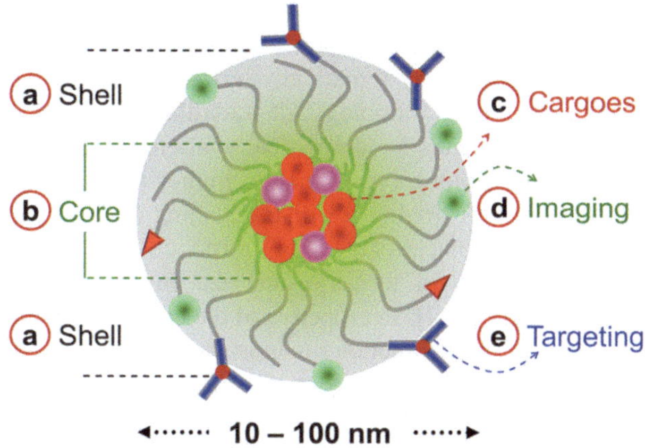

Fig. 1 Schematic representation of the concept of an 'idealised' spherical self-assembled multifunctional carrier composed of hydrophilic shell (*a*) and hydrophobic core (*b*), loaded with wide varieties of therapeutic cargoes (*c*). Additional functionalities such as imaging agents (*d*) and/or (*e*) targeting agents such as antibodies or small-molecule ligands can be integrated onto the nanocarrier, to enhance the versatility and applicability of the carrier

polymerisation (ATRP) [16] and reversible addition-fragmentation chain transfer (RAFT) [17, 18] polymerisation have been developed. These polymerisation processes result in nondegradable polymers, and this could severely limit the scope and applicability of these techniques in nanomedicine. Innovative approaches involving the use of cyclic ketene acetals as comonomers have to some extent addressed this limitation by introducing degradable ester bonds along the backbone [19]. From the toxicity and safety perspective, it is highly desirable to have versatile synthetic access to functional degradable polymers. In this regard ring-opening polymerisation (ROP) of cyclic monomers has emerged as a powerful synthetic tool. Starting from respective cyclic monomers (such as lactides, lactones, *N*-carboxy anhydrides, carbonates and phosphoesters), different classes of degradable polymer can be readily accessed [20]. Recently significant advances have been made in developing facile synthetic access to functional cyclic monomers that significantly expanded the applicability of these degradable polymers [21, 22]. With the advent of transition metal-free organocatalysis, degradable materials with excellent biocompatibility can be accessed as these catalysts eliminate toxicity concerns that arise due to trace impurities from traditional metal-based catalysts [23]. Efforts have been directed towards development of chain transfer agents and initiators to effect controlled polymerisation of mechanistically distinct classes of polymers to bring in added functionalities [24]. With the combination of these methodologies, critical molecular variables including molecular weight, molecular weight distribution, chemical composition, functionalities, chemistries of α- and ω-chain ends, relative sizes of the blocks, block sequences, topologies, degradability and physical properties can be tailored. By having access to vast 'library' of well-defined amphiphilic building blocks, the chemical and physical nature of the resultant nanostructural ensemble can be systematically modified (Fig. 2). For instance, by moving from a linear to a cyclic topology, degradation behaviour, biodistribution and PK of these materials can be dramatically altered [25, 26]. Similarly by extending from a simple diblock copolymer to higher block configurations, one could effectively tap on their qualitative distinguishing features and richer phase diagrams that are otherwise inaccessible [11]. Introduction of branching will have impact on flexibility, which in turn impacts the in vivo behaviour such as biodistribution and clearance [27]. Irrespective of the composition and architecture, the minimalistic design of multifunctional nanoscale self-assembled soft materials for biomedical delivery application consists of at least one each of hydrophilic and hydrophobic segment, a simple diblock copolymer. In the following subsection, briefly, the role and function of each of these segments will be described.

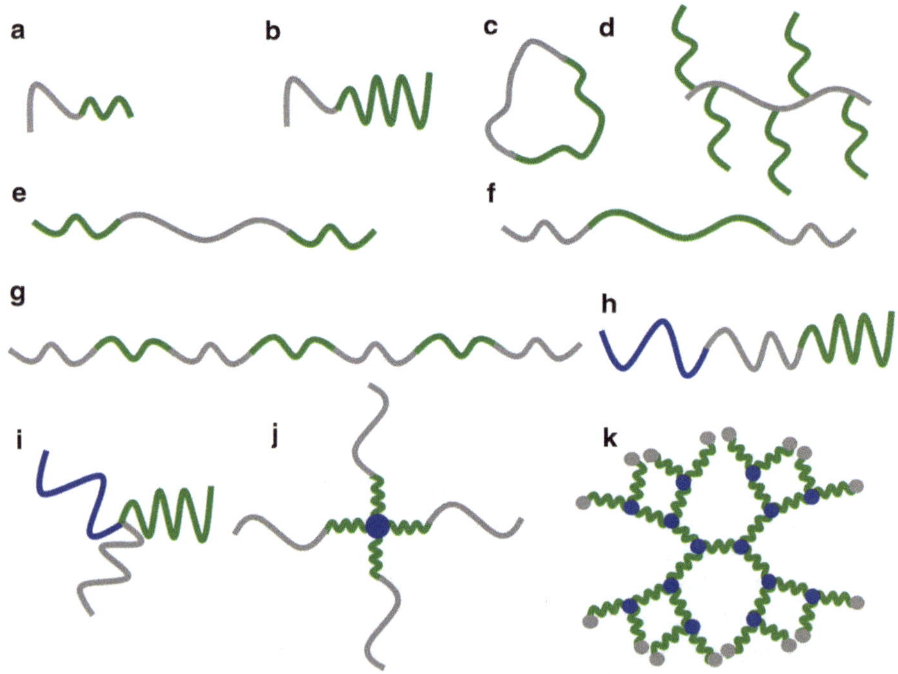

Fig. 2 Schematic representation of the readily accessible well-defined polymeric amphiphilic architectures such as AB diblock copolymers of different block compositions (**a** and **b**), cyclic block copolymer (**c**), graft copolymer (**d**), ABA triblock copolymers with varied block sequence (**e** and **f**), multiblock copolymers (**g**), ABC triblock copolymers (**h**), miktoarm star and dendritic/hyperbranched polymers (**i–k**, respectively). Collectively, these selected examples illustrate the synthetic versatility of accessible polymer architectures

2.1.1 Hydrophilic Segment

Hydrophilic materials serve as the interface between the carrier (with the drug) and the biological components. The actual choice of hydrophilic polymer will impact the chemical nature of the hydrophilic shell and its interaction with proteins, which in turn would control the cascade of biological responses, playing a crucial role in many processes. In nanoparticle-based drug delivery, the very nature of protein-binding events can influence the immunological response and pharmacodynamic properties and eventually contribute towards clearance from body. As numerous biological cascade events are triggered by the nature of proteins adsorbed/bound onto the surface, that in turn can affect the nature of immunoresponse.

Hydrophilic materials can be designed either as electrostatically charged (positive or negative) or neutral. Typically PEG is preferred as the hydrophilic part as it is an FDA-approved component. PEG brings in water solubility and impart 'stealth' – to be not detected by the immune system and hence prolongs the blood circulation. Many PEG derivatives are also commercially available and are cheap. In spite of all these positive attributes of PEG, its non-biodegradability, hypersensitivity and toxicity of its side products have necessitated to look for hydrophilic neutral PEG alternatives [28, 29]. Amongst the available alternative options, zwitterionic polymers have emerged as a promising candidate [30], as these polymers demonstrate excellent resistance to non-specific protein binding [31]. Moreover, the concept of zwitterion can be integrated in many platform materials including degradable polymers with other functionalities, rendering this approach very versatile to generate hydrophilic components.

Apart from neutral, ionic surfaces such as cationic or anionic shell can also be designed. Unlike PEG shell, these surfaces pack in reactive handles that can be used for additional reactions. For instance, the shell region of the micelles can be cross-linked to further enhance the stability of these micelles [32]. Such an introduction of cross-linking in the shell region improves the stability of the nanoparticles, particularly under in vivo conditions where typically injected nanoparticles in the bloodstream undergoes infinite dilution that could lead to premature disassembly (Fig. 3a). Apart from enhancing the stability via cross-linking, the chemical functionalities such as carboxylic acid or amino groups can be used to conjugate additional functionalities (such as imaging agents or targeting moieties), paving access to multifunctional nanocarriers [33].

2.1.2 Hydrophobic Segment

For serious clinical translation and commercialisation, it is desirable to develop chemistries based on biodegradable polymers that are not toxic to the host and are easily absorbed and processed by the body [34]. Compared to monomers for nondegradable polymers (styrenics and (meth)acrylate), commercially available monomer options to access degradable polymers (lactides, caprolactone,

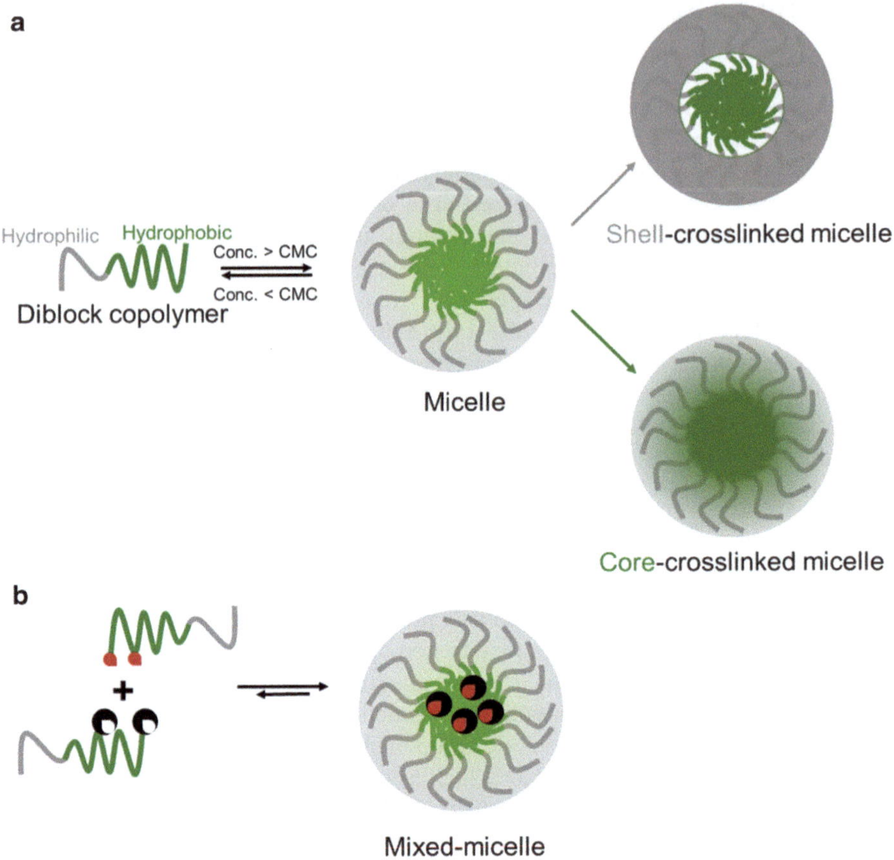

Fig. 3 Schematic representation of different strategies to enhance the micellar stability. (**a**) The shell region or the core region of micelles can be covalently cross-linked to ensure that the micelles do not disassociate under high dilution. (**b**) Alternatively, complementary supramolecular moieties can be integrated to form mixed micelles that can contribute in the enhancement of stability via molecular recognition events

trimethylene carbonate) are in general limited. Recognising the opportunities with development of synthetic access to well-defined degradable polymers, lots of efforts have been directed in development of various classes of degradable polymers (such as polyphosphoesters [35], polyesters [36], polycarbonates [37], polypeptides [38]) that could serve as the hydrophobic segment. With these developments, functionalities comparable to that of styrenics and (meth)acrylate can be introduced onto degradable polymers.

Since one key aspect of the core is to serve as a host for water-insoluble drug, the ability to tailor the hydrophobic composition can enhance drug-loading capacity of the nanocarrier [39]. Similar to shell cross-linking approaches, core cross-linking (Fig. 3a) via covalent chemistry [40] and non-covalent interactions assisted with mixed-micellisation-based [41, 42] (Fig. 3b) approaches are identified as strategies to enhance the stability of the nanostructures in harsh and demanding physiological conditions. With the judicious combination of hydrophobicity and molecular recognition units, the shape of nanostructures can also be modulated [43]. The ability to engineer the core chemistry effectively, tailored for specific drug, is crucial to ensure the clinical and commercial success of products [44].

2.2 Self-Assembly: Possibilities with Size and Shape Control

Though access to nanostructures by molecular self-assembly route is simple and spontaneous [5], predictable and reproducible control of nanostructural shape nanostructures has been a challenge, particularly with the innumerable possibilities in the design of amphiphiles that can span over 2–3 orders of magnitude in their molecular weights [11]. Traditionally 'packing parameter concept' is used to rationalise the observed shapes [45]. Packing parameter, p, is defined by the following equation: $p = v/(al)$, where v is the volume of the hydrophobic tail, l is the length of the hydrophobic tail, and a corresponds to interfacial area per molecule. Depending upon nanostructural shape, p will vary significantly. Typically spherical micelles will have p values less than 1/3. Elongated micelles will have an intermediate p values between 1/3 and 1/2, whereas p values for bilayers range from 1/2 to 1. Though this concept is useful in understanding the observed morphologies, it significantly suffers from lack of predictive power. This limitation primarily stems from the fact that some of the variables considered in calculating p are in fact are thermodynamic properties that cannot be simply estimated based on geometric consideration of chemical structures [46]. Alternatively, for block copolymers, ratio of hydrophobic to hydrophilic components have also been used in literature to track the changes in the morphology of resultant nanostructures [47, 48]. However, this approach might not be perfect either, owing to inherent differences in the nature and strength of interactions across different classes of polymers and thereby leading to significant differences. Though with the recent advances in theoretical methods such as self-consistent field theory (SCFT) and computational power, prediction of polymer self-assembly has been feasible [10]. It should still be pointed out that solution-state self-assembly with complex polymeric systems can be challenging as the self-assembly behaviour can be influenced by numerous chemical and physical parameters (e.g. pH, temperature, salt concentrations). In spite of these practical limitations in predictive power, tremendous progress has been made in the development of self-assembled materials with precise shape and size control.

Synthetic ability to access complex and yet precise compositions, coupled with advances in the theoretical understanding of underlying physics of self-assembly process, has translated to control over the nanostructural morphology. Amongst accessible nanostructural shapes, only a few shapes such as spherical micelles, elongated micelles, vesicles and disc-like micelles have emerged as promising candidates for drug delivery applications [49] (Fig. 4). Exotic shapes such as toroids

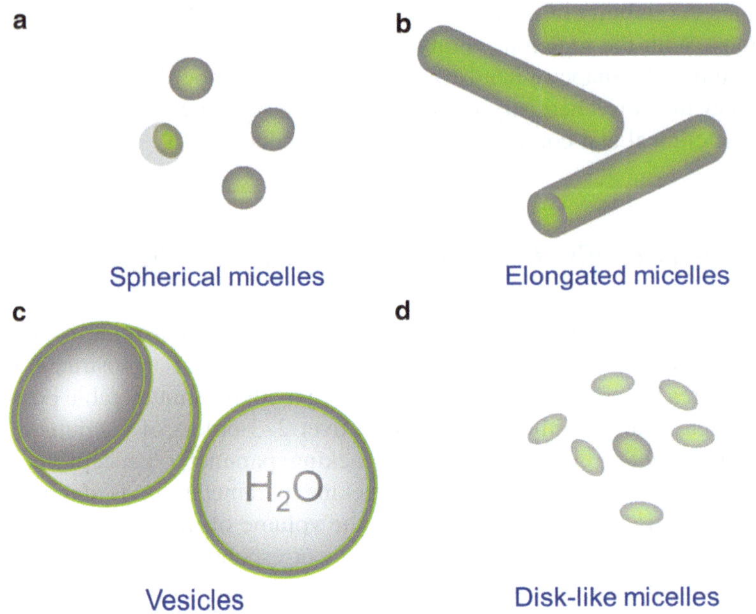

Fig. 4 Schematic representation of the commonly explored self-assembled nanostructural shapes such as spherical micelles (**a**), elongated rod-like micelles or worm-like micelles (**b**) vesicles (**c**) and disc-like micelles (**d**)

[50], helical micelles [51], hamburger micelles [52], stacked discs [53], etc., though exciting, may not be promising for clinical translation. Amongst the accessible morphologies, spherical micelles constitute the simplest and well-studied. Owing to their relative simplicity in design and easy accessibility, currently there are many spherical block copolymeric micellar nanocarriers at preclinical or clinical evaluation [54–58]. These systems are typically core-shell micellar nanostructures, primarily designed to enhance therapeutic efficacy and minimise adverse side effects.

In the light of exciting reports that demonstrate the role of nanostructural shape in cellular internalisation [59] and prolonged blood circulation [60], nanostructural shape control is emerging as an important strategy to enhance the therapeutic outcome. Also, the nature of self-assembled morphology can dictate the scope of application. For instance, vesicles with hydrophilic internal cavity can serve as a reservoir for hydrophilic components, whereas in the hydrophobic interface it would serve as a barrier for the rapid release of the hydrophilic cargoes. At the same time the hydrophobic interface can also be used for loading hydrophobic drugs, enabling co-delivery of multiple therapeutics with distinct chemical nature [8]. Compared to vesicles made from small-molecule amphiphiles (lipids), vesicles made from block copolymers have been shown to have higher stability and elasticity [11, 61]. Moreover, additional functionalities such as degradability can be easily integrated into the chemistry of vesicle-forming block copolymer, enabling control over the rate of cargo release [62]. Disc-like micelles due to their unique aspect ratio and their strongly segregated core would also be interesting as a nanocarrier for drug delivery applications [63].

3 Advantages with the Use of Nanocarriers

Since the very first FDA approval of nanocarrier (liposome-encapsulated doxorubicin [64]), there are many 'first-generation' nanoformulations that are either approved or in the various stages of clinical trials [65]. Nanocarriers have empowered formulation specialist to (re)visit potent drugs that were considered undevelopable [4, 66]. Reduction in side effects (in comparison with free drug), improved drug half-life and solubility of poorly water-soluble drugs were some of the key value addition, brought in with the nanocarriers. As with the anticancer therapeutics, the ability to control the size and numerous other properties of nanocarriers enabled passive accumulation in tumour via enhanced permeation and retention (EPR) effect [67]. As for the delivery of complex biomacromolecules such as proteins, *si*RNA and DNA, these nanocarriers can be designed to navigate harsh in vivo conditions, wherein without the 'protection' of nanocarriers, these biologics would be degraded. Effective engineering of intracellular trafficking of therapeutic nanocarriers can have major impact combating drug resistance. In contrast to conventional pharmaceutics, functional carriers can be also be incorporated with stimuli-responsive components that can respond to differences in microenvironment thermal- and/or pH-based stimuli [68]. In line with the vision of multifunctional nanocarriers (Fig. 1), the pursuit of carriers integrated with active-targeting abilities and built-in imaging agents could in principle revolutionise the way patients are treated. In spite of ongoing debates on the need and merits/ demerits of some of these combinations, these research efforts paved a solid foundation for future exciting applications [69].

4 Challenges

Incredible possibilities offered by the nanoformulation in therapeutic delivery and imaging systems also pose unique challenges related to their clinical translation [70]. Most of the approved nanoformulated therapeutics currently used in clinic have so far have relied just on mitigating toxicity (compared to the parent drug) and not on improving the overall efficacy [71]. Though reducing the side effects is desirable, high costs often associated with these nanomedicines have demanded critical evaluation of current nanomedicines. Also compared to conventional pharmaceuticals, clinical development path for nanoformulations is relatively difficult. Strategies to systematically combat these challenges should be in place, even at the preclinical development to ensure that the lead formulations do not encounter issues at the late stages in the development [72]. In this section, some of the key challenges in the development of nanomedicine are highlighted.

4.1 Need for Rigorous Characterisation

Exact composition and purity of API in conventional pharmaceutics can be precisely determined by using one or combination of readily accessible techniques such as nuclear magnetic resonance (NMR) spectroscopy, high-pressure liquid chromatography (HPLC), elemental analysis, X-ray crystallography and mass spectrometry. Similarly batch-to-batch quality control can also be effectively implemented. In comparison with these well-defined molecular APIs, nanoparticulate drug delivery systems are ensemble of multiple components. Hence their characterisation is significantly more challenging and also distinct from that of conventional pharmaceuticals. Several parameters including the chemical composition, particle size, shape, polydispersity, surface charge, charge density, surface ligands, ligand density, stability and the nature of interactions of these nanoparticles with other components in the formulation are considered crucial for the biomedical applications [73]. Since many of these properties are interlinked in complex way, any minor changes with one specific parameter would lead to unpredictable pharmacokinetics and toxicity [74]. Hence, in-depth physico-chemical characterisation of these nanoparticles is crucial to ensure not only homogeneity and batch-to-batch consistency but also overall safety [75, 76].

To highlight the complexity of characterising nanocarriers, determination of their size will be used as an illustrative example. Size of nanostructure depending upon the actual measurement technique could vary [75]. Dynamic light scattering (DLS) is commonly used to determine hydrodynamic diameter. In DLS, an autocorrelation function generated by the intensity fluctuations of the scattered light (due to Brownian motion of nanoparticles) is used to obtain translational diffusion coefficient, which in turn is used to determine the hydrodynamic diameter by using Stokes–Einstein equation [77]. The raw data (autocorrelation function) are fitted with calculations based on assumed distributions to obtain the hydrodynamic diameter. The simplicity of sample preparation for DLS makes this technique useful. The size of nanostructures can be determined in the conditions that would realistically reflect the actual environment in which the nanoparticles are actually used, such as temperature, pH, salt concentrations and the presence of other additives. Though DLS is a readily accessible technique, providing rapid and reproducible estimate of particle size and polydispersity index, there are few inherent limitations with this technique. Characterisation of nonspherical, polydisperse, concentrated and coloured samples (interfering with the laser wavelength) by DLS is challenging. For instance, in polydisperse sample, scattered intensity from a minor fraction of particles with larger diameter could almost entirely mask the contributions from a major population of smaller particles, leading to overestimation of size. Also depending upon the actual statistical treatment of the data (intensity, volume or number average), the diameter values could differ significantly. To overcome these limitations of DLS, often nanoparticles are characterised by complementary microscopy techniques such as atomic force microscopy (AFM) and transmission electron microscopy (TEM). As these microscopy techniques rely on different principles and sample state could also be different, resultant data must be interpreted by bearing these differences in mind. For example, diameters obtained from DLS are often larger than TEM data. This difference could be primarily due to the inherent differences in the sample state in these respective techniques. Unlike DLS, in conventional TEM, sample is typically maintained under high vacuum (dry state) on a supported grid, which could contribute towards shrinkage. TEM in general can only provide information on 'cross sections'. In order to obtain 3-D information on morphologies of complex nanostructures, tomography or other complementary characterisation should be performed. The limitations with conventional TEM such as the need to use staining agents generate contrast so as to visualise soft matter, and also the possibility of introduction of artefacts due to drying of samples can be eliminated with the use of cryogenic transmission electron microscopy (cryo-TEM), wherein the sample is essentially preserved in their native state in a thin vitrified ice sheet [78]. Though cryo-TEM is a powerful tool, it does need additional investment on hardware accessories and highly trained technicians to acquire data. Selection of a particular characterisation technique amongst multitude of option should be essentially done by taking into consideration of inherent strengths and limitation of the technique also with other factors such as cost and sample preparation conditions [73].

Apart from the measurement technique, the actual sample environment can also affect the data [76]. Size of drug-loaded micelles in solution, in equilibrium with some of the free unencapsulated drug, could be different from that of drug-loaded micelles without any free drugs and unloaded micelles. Also, presence of serum protein in the dispersing medium could contribute in changes in sizes due to the interaction of these proteins with the hydrophilic surface of micelles. Similarly, thermal history of the sample, changes in ionic strength and pH of the dispersing medium can all have drastic influence on the size as well. Since the size of nanoparticle partly encodes the mechanism of cellular uptake and the uptake pathway, it is very important to take note of these factors into consideration. For insightful coverage of the challenges associated with entire array of characterisation in the nanomedicines, readers may refer to the following publications [76, 73]. Recognising thc importance of thorough characterisation, protocols are being developed by taking into consideration specific challenges and also ensuring that these protocols are standardised so that measurement across different sites and even across different classes of samples can be compared [72].

4.2 Toxicity Considerations

Systematic and quantitative understanding of pharmacokinetics of nanoformulations would contribute towards better understanding of the biodistribution and clearance of nanostructures. This fundamental understanding forms the basis for clear understanding of toxicity of nanocarriers. The very unique physical and chemical attributes of nanomaterials that render them attractive for the biomedical applications do bring in challenges due to unpredictable interactions with the biological systems [74]. Unlike scenarios dealing with accidental exposure, with nanocarriers, engineered nanomaterials are deliberately introduced into the body. Criteria for deciding on acceptable toxicity levels should be based on the frequency of administration. For instance, diagnostic use may not require frequent administration when compared to the usage as a carrier for drugs that require administration over a long period of time, and hence immunological considerations such as hypersensitivity due to complement activation would be very different, respectively [79]. Thorough understanding of the metabolic path of every component used in the formulation is necessary. As with the nondegradable components, efforts should be directed towards understanding the pathways of elimination of these materials from the body. As for the degradable materials, understanding of the metabolic path is crucial. Spatio-temporal implications of degradation should be carefully considered. There would be stark difference in intracellular degradation of polymeric micelles versus disintegration of micelles in the bloodstream. Also, the toxicity profiles of the degraded components should be assessed. Whilst assessing toxicity, even trace impurities in starting materials should be taken into consideration. For instance, traces of catalytic components could be toxic even at ppm levels. Though integration of transition elements and metallic components could bring in additional exciting attributes such as imaging capabilities, their element-specific toxicity attributes and ability to generate reactive oxygen species (ROS) must be carefully considered [80].

As the risks associated with the exposure to nanomaterials are yet to be fully understood, critical evaluation of toxicological aspects of nanomedicine is necessary. The entire gamut of nanomaterials is in a relatively early stage of deployment, and comprehensive understanding of interplay of chemical composition and physical attributes (such as size, shape, surface charges) with the biological systems is still lacking [79]. This is because of the complexity of toxicological characterisation of nanoformulation due to dynamic nature of these samples and the stark differences in spatio-temporal interactions across multiple levels (whole body, organs, tissue, cellular, subcellular level) [81]. Interaction of nanocarrier with biological components could alter several attributes of the carrier. For instance, changes in the surface composition and size of nanoparticles due to protein binding in the bloodstream would make the toxicity prediction difficult [82]. Hence in-depth in vitro and in vivo studies are necessary to ensure that exposure to these materials is safe. To address some of the major toxicity concerns arising from immunology perspective, detailed and appropriately designed in vivo studies are crucial [81]. For instance, in the tumour models the aggressiveness of tumours is dependent on the actual animal model. Care should be taken to ensure that simplistic models are not chosen to demonstrate 'success'. With the deployment of cytotoxic API-loaded nanocarriers (anticancer agents), biodistribution studies should be carefully evaluated, as often a significant portion of the drugs are offloaded to healthy organs, leading to serious side effects. Characterisation of reaction of body's immune system to nanocarrier cannot be captured fully in a simple in vitro experiment. It is important that the cell lines and in vivo models used are relevant to humans and also must also be a better representation of realistic scenarios [76].

5 Conclusion and Outlook

Nanomedicines provide exciting opportunities to improve upon our arsenal of treatment options against numerous challenging diseases. It has incredible potential for revolutionising the therapeutics and diagnostics. The developments in synthetic chemistry have enabled unprecedented access to well-defined functional building blocks that can be tailored for the specific application, with almost no limitation as to the class of therapeutics that one can potentially work with (such as hydrophobic small molecules, proteins, *si*RNA, gene, etc.). Compared to conventional pharmaceuticals, these nanocarrier-based drug deliveries demand collaborative and multidisciplinary approach to have better understanding of the dynamic physicochemical properties and complex spatio-temporal biological interactions, and for successful clinical translation, in-depth characterisation is crucial. Also, owing to regulatory hurdles, the competition that typically arises with the key patent expiry from generic version manufacturers could be rather limited with the nanoformulations [64, 83]. Hence, a well-characterised, safe and effective nanocarrier-based formulation will not only have positive impact on patients but also on the pharmaceutical industry business model.

Acknowledgements This work was funded by the Institute of Bioengineering and Nanotechnology (Biomedical Research Council, Agency for Science, Technology and Research, Singapore). Dr. Yi Yan Yang's support for this work is greatly appreciated.

References

1. Elebring T, Gill A, Plowright AT (2012) What is the most important approach in current drug discovery: doing the right things or doing things right? Drug Discov Today 17 (21–22):1166–1169
2. Paul SM, Mytelka DS, Dunwiddie CT, Persinger CC, Munos BH, Lindborg SR, Schacht AL (2010) How to improve R&D productivity: the pharmaceutical industry's grand challenge. Nat Rev Drug Discov 9(3):203–214

3. Park K (2007) Nanotechnology: what it can do for drug delivery. J Control Release 120 (1–2):1–3
4. Shi J, Votruba AR, Farokhzad OC, Langer R (2010) Nanotechnology in drug delivery and tissue engineering: from discovery to applications. Nano Lett 10(9):3223–3230
5. Whitesides GM, Boncheva M (2002) Beyond molecules: Self-assembly of mesoscopic and macroscopic components. Proc Natl Acad Sci U S A 99:4769–4974
6. Deshayes S, Kasko AM (2013) Polymeric biomaterials with engineered degradation. J Polym Sci A Polym Chem 51(17):3531–3566
7. Göpferich A (1996) Mechanisms of polymer degradation and erosion. Biomaterials 17 (2):103–114
8. Blanazs A, Armes SP, Ryan AJ (2009) Self-assembled block copolymer aggregates: from micelles to vesicles and their biological applications. Macromol Rapid Commun 30 (4–5):267–277
9. Kataoka K, Harada A, Nagasaki Y (2001) Block copolymer micelles for drug delivery: design, characterization and biological significance. Adv Drug Deliv Rev 47(1):113–131
10. Bates FS, Hillmyer MA, Lodge TP, Bates CM, Delaney KT, Fredrickson GH (2012) Multiblock polymers: panacea or pandora's box? Science 336(6080):434–440
11. Halperin A (2006) Polymeric *vs.* monomeric amphiphiles: design parameters. Polym Rev 46:173–214
12. Torchilin VP (2012) Multifunctional nanocarriers. Adv Drug Delivery Rev 64 (Suppl):302–315
13. Iha RK, Wooley KL, Nyström AM, Burke DJ, Kade MJ, Hawker CJ (2009) Applications of orthogonal "click" chemistries in the synthesis of functional soft materials. Chem Rev 109:5620–5686
14. Braunecker WA, Matyjaszewski K (2007) Controlled/living radical polymerization: features, developments, and perspectives. Prog Polym Sci 32:93–146
15. Hawker CJ, Bosman AW, Harth E (2001) New polymer synthesis by nitroxide mediated living radical polymerizations. Chem Rev 101:3661–3688
16. Matyjaszewski K, Xia J (2001) Atom transfer radical polymerization. Chem Rev 101:2921–2990
17. Moad G, Rizzardo E, Thang SH (2008) Toward living radical polymerization. Acc Chem Res 41:1133–1142
18. Perrier S, Takolpuckdee P (2005) Macromolecular design via reversible addition–fragmentation chain transfer (RAFT)/xanthates (MADIX) polymerization. J Polym Sci A Polym Chem 43(22):5347–5393
19. Agarwal S (2010) Chemistry, chances and limitations of the radical ring-opening polymerization of cyclic ketene acetals for the synthesis of degradable polyesters. Polym Chem 1 (7):953–964
20. Nuyken O, Pask SD (2013) Ring-opening polymerization – an introductory review. Polymers 5:361–403
21. Pratt R, Nederberg F, Waymouth RM, Hedrick JL (2008) Tagging alcohols with cyclic carbonate: a versatile equivalent of (meth)acrylate for ring-opening polymerization. Chem Commun 114–116
22. Sanders DP, Fukushima K, Coady DJ, Nelson A, Fujiwara M, Yasumoto M, Hedrick JL (2010) A simple and efficient synthesis of functionalized cyclic carbonate monomers using a versatile pentafluorophenyl ester intermediate. J Am Chem Soc 132:14724–14726
23. Kiesewetter MK, Shin EJ, Hedrick JL, Waymouth RM (2010) Organocatalysis: opportunities and challenges for polymer synthesis. Macromolecules 43:2093–2107
24. Bernaerts KV, Du Prez FE (2006) Dual/heterofunctional initiators for the combination of mechanistically distinct polymerization techniques. Prog Polym Sci 31(8):671–722
25. Nasongkla N, Chen B, Macaraeg N, Fox ME, Fréchet JMJ, Szoka FC (2009) Dependence of pharmacokinetics and biodistribution on polymer architecture: effect of cyclic versus linear polymers. J Am Chem Soc 131(11):3842–3843

26. Hoskins JN, Grayson SM (2009) Synthesis and degradation behavior of cyclic poly(ε-caprolactone). Macromolecules 42(17):6406–6413
27. Fox ME, Szoka FC, Fréchet JMJ (2009) Soluble polymer carriers for the treatment of cancer: the importance of molecular architecture. Acc Chem Res 42:1141–1151
28. Barz M, Luxenhofer R, Zentel R, Vicent MJ (2011) Overcoming the PEG-addiction: well-defined alternatives to PEG, from structure–property relationships to better defined therapeutics. Polym Chem 2(9):1900–1918
29. Knop K, Hoogenboom R, Fischer D, Schubert US (2010) Poly(ethylene glycol) in drug delivery: pros and cons as well as potential alternatives. Angew Chem Int Ed 49 (36):6288–6308
30. Jiang S, Cao Z (2010) Ultralow-fouling, functionalizable, and hydrolyzable zwitterionic materials and their derivatives for biological applications. Adv Mater 22(9):920–932
31. Ladd J, Zhang Z, Chen S, Hower JC, Jiang S (2008) Zwitterionic polymers exhibiting high resistance to nonspecific protein adsorption from human serum and plasma. Biomacromolecules 9(5):1357–1361
32. O'Reilly RK, Hawker CJ, Wooley KL (2006) Cross-linked block copolymer micelles: functional nanostructures of great potential and versatility. Chem Soc Rev 35:1068–1083
33. Elsabahy M, Wooley KL (2012) Design of polymeric nanoparticles for biomedical delivery applications. Chem Soc Rev 41(7):2545–2561
34. Tian H, Tang Z, Zhuang X, Chen X, Jing X (2012) Biodegradable synthetic polymers: Preparation, functionalization and biomedical application. Prog Polym Sci 37(2):237–280
35. Wang Y-C, Yuan Y-Y, Du J-Z, Yang X-Z, Wang J (2009) Recent progress in polyphosphoesters: from controlled synthesis to biomedical applications. Macromol Biosci 9 (12):1154–1164
36. Seyednejad H, Ghassemi AH, van Nostrum CF, Vermonden T, Hennink WE (2011) Functional aliphatic polyesters for biomedical and pharmaceutical applications. J Control Release 152 (1):168–176
37. Feng J, Zhuo R-X, Zhang X-Z (2012) Construction of functional aliphatic polycarbonates for biomedical applications. Prog Polym Sci 37(2):211–236
38. Cheng J, Deming T (2012) Synthesis of polypeptides by ring-opening polymerization of α-amino acid N-carboxyanhydrides. In: Deming T (ed) Peptide-based materials, vol 310. Topics in Current Chemistry. Springer, Berlin Heidelberg, pp 1–26
39. Lee ALZ, Venkataraman S, Sirat SBM, Gao S, Hedrick JL, Yang YY (2012) The use of cholesterol-containing biodegradable block copolymers to exploit hydrophobic interactions for the delivery of anticancer drugs. Biomaterials 33(6):1921–1928
40. Wiltshire JT, Qiao GG (2007) Recent advances in star polymer design: degradability and the potential for drug delivery. Aust J Chem 60(10):699–705
41. Yang C, Ebrahim Attia AB, Tan JPK, Ke X, Gao S, Hedrick JL, Yang Y-Y (2012) The role of non-covalent interactions in anticancer drug loading and kinetic stability of polymeric micelles. Biomaterials 33(10):2971–2979
42. Ebrahim Attia AB, Ong ZY, Hedrick JL, Lee PP, Ee PLR, Hammond PT, Yang Y-Y (2011) Mixed micelles self-assembled from block copolymers for drug delivery. Curr Opin Colloid Interface Sci 16(3):182–194
43. Venkataraman S, Chowdhury ZA, Lee AL, Tong YW, Akiba I, Yang YY (2013) Access to different nanostructures via self-assembly of thiourea-containing PEGylated amphiphiles. Macromol Rapid Commun 34(8):652–658
44. Ding J, Chen L, Xiao C, Chen L, Zhuang X, Chen X (2014) Noncovalent interaction-assisted polymeric micelles for controlled drug delivery. Chem Commun 50:11274–11290
45. Israelachivili J, Mitchell DJ, Ninham BW (1976) Theory of self-assembly of hydrocarbon amphiphiles into micelles and bilayers. J Chem Soc Faraday Trans 2(72):1525–1568
46. Nagarajan R (2002) Molecular packing parameter and surfactant self-assembly: the neglected role of the surfactant tail. Langmuir 18:31–38

47. Zupancich JA, Bates FS, Hillmyer MA (2006) Aqueous dispersions of poly(ethylene oxide)-*b*-poly(γ-methyl-μ-caprolactone) block copolymers. Macromolecules 39:4286–4288
48. Venkataraman S, Hedrick JL, Yang YY (2014) Fluorene-functionalized aliphatic polycarbonates: design, synthesis and aqueous self-assembly of amphiphilic block copolymers. Polym Chem 5(6):2035–2040
49. Venkataraman S, Hedrick JL, Ong ZY, Yang C, Ee PLR, Hammond PT, Yang YY (2011) The effects of polymeric nanostructure shape on drug delivery. Adv Drug Deliv Rev 63 (14–15):1228–1246
50. Pochan DJ, Chen Z, Cui H, Hales K, Qi K, Wooley KL (2004) Toroidal triblock copolymer assemblies. Science 306:94–97
51. Zhong S, Cui H, Chen Z, Wooley KL, Pochan DJ (2008) Helix self-assembly through the coiling of cylindrical micelles. Soft Matter 4:90–93
52. Li Z, Hillmyer MA, Lodge TP (2006) Morphologies of multicompartment micelles formed by ABC miktoarm star terpolymers. Langmuir 22:9409–9417
53. Venkataraman S, Lee AL, Maune HT, Hedrick JL, Prabhu VM, Yang YY (2013) Formation of Disk- and Stacked-Disk-like Self-Assembled Morphologies from Cholesterol-Functionalized Amphiphilic Polycarbonate Diblock Copolymers. Macromolecules 46(12):4839–4846
54. Hrkach J, Von Hoff D, Ali MM, Andrianova E, Auer J, Campbell T, De Witt D, Figa M, Figueiredo M, Horhota A, Low S, McDonnell K, Peeke E, Retnarajan B, Sabnis A, Schnipper E, Song JJ, Song YH, Summa J, Tompsett D, Troiano G, Van Geen Hoven T, Wright J, LoRusso P, Kantoff PW, Bander NH, Sweeney C, Farokhzad OC, Langer R, Zale S (2012) Preclinical development and clinical translation of a PSMA-targeted docetaxel nanoparticle with a differentiated pharmacological profile. Sci Trans Med 4(128):128–139
55. Kato K, Chin K, Yoshikawa T, Yamaguchi K, Tsuji Y, Esaki T, Sakai K, Kimura M, Hamaguchi T, Shimada Y, Matsumura Y, Ikeda R (2012) Phase II study of NK105, a paclitaxel-incorporating micellar nanoparticle, for previously treated advanced or recurrent gastric cancer. Invest New Drugs 30(4):1621–1627
56. Kieler-Ferguson HM, Fréchet JMJ, Szoka FC Jr (2013) Clinical developments of chemotherapeutic nanomedicines: polymers and liposomes for delivery of camptothecins and platinum (II) drugs. Nanomed Nanobiotechnol 5(2):130–138
57. Endo K, Ueno T, Kondo S, Wakisaka N, Murono S, Ito M, Kataoka K, Kato Y, Yoshizaki T (2013) Tumor-targeted chemotherapy with the nanopolymer-based drug NC-6004 for oral squamous cell carcinoma. Cancer Sci 104(3):369–374
58. Yamamoto Y, Hyodo I, Takigahira M, Koga Y, Yasunaga M, Harada M, Hayashi T, Kato Y, Matsumura Y (2014) Effect of combined treatment with the epirubicin-incorporating micelles (NC-6300) and 1,2-diaminocyclohexane platinum (II)-incorporating micelles (NC-4016) on a human gastric cancer model. Int J Cancer 135(1):214–223
59. Champion JA, Mitragotri S (2006) Role of target geometry in phagocytosis. Proc Natl Acad Sci U S A 103:4930–4934
60. Geng Y, Dalhaimer P, Cai S, Tsai R, Tewari M, Minko T, Discher DE (2007) Shape effects of filaments versus spherical particles in flow and drug delivery. Nat Nanotechnol 2:249–255
61. Discher BM, Won Y-Y, Ege DS, Lee JCM, Bates FS, Discher DE, Hammer DA (1999) Polymersomes: tough vesicles made from diblock copolymers. Science 284:1143–1146
62. Ahmed F, Pakunlu RI, Brannan A, Bates F, Minko T, Discher DE (2006) Biodegradable polymersomes loaded with both paclitaxel and doxorubicin permeate and shrink tumors, inducing apoptosis in proportion to accumulated drug. J Control Release 116(2):150–158
63. Holder SJ, Sommerdijk NAJM (2011) New micellar morphologies from amphiphilic block copolymers: disks, toroids and bicontinuous micelles. Polym Chem 2(5):1018–1028
64. Barenholz Y (2012) Doxil® — the first FDA-approved nano-drug: lessons learned. J Control Release 160(2):117–134
65. Davis ME, Chen Z, Shin DM (2008) Nanoparticle therapeutics: an emerging treatment modality for cancer. Nat Rev Drug Discov 7(9):771–782

66. Cooper ER (2010) Nanoparticles: a personal experience for formulating poorly water soluble drugs. J Control Release 141(3):300–302
67. Maeda H, Nakamura H, Fang J (2013) The EPR effect for macromolecular drug delivery to solid tumors: Improvement of tumor uptake, lowering of systemic toxicity, and distinct tumor imaging in vivo. Adv Drug Deliv Rev 65(1):71–79
68. Danhier F, Feron O, Préat V (2010) To exploit the tumor microenvironment: passive and active tumor targeting of nanocarriers for anti-cancer drug delivery. J Control Release 148 (2):135–146
69. Cheng Z, Al Zaki A, Hui JZ, Muzykantov VR, Tsourkas A (2012) Multifunctional nanoparticles: cost versus benefit of adding targeting and imaging capabilities. Science 338 (6109):903–910
70. Wei A, Mehtala JG, Patri AK (2012) Challenges and opportunities in the advancement of nanomedicines. J Control Release 164(2):236–246
71. Venditto VJ, Szoka FC Jr (2013) Cancer nanomedicines: so many papers and so few drugs! Adv Drug Deliv Rev 65(1):80–88
72. Zamboni WC, Torchilin V, Patri AK, Hrkach J, Stern S, Lee R, Nel A, Panaro NJ, Grodzinski P (2012) Best practices in cancer nanotechnology: perspective from NCI nanotechnology alliance. Clin Cancer Res 18(12):3229–3241
73. Lin P-C, Lin S, Wang PC, Sridhar R (2014) Techniques for physicochemical characterization of nanomaterials. Biotechnol Adv 32(4):711–726
74. Kagan VE, Bayir H, Shvedova AA (2005) Nanomedicine and nanotoxicology: two sides of the same coin. Nanomed Nanotechnol Biol Med 1(4):313–316
75. Crist RM, Grossman JH, Patri AK, Stern ST, Dobrovolskaia MA, Adiseshaiah PP, Clogston JD, McNeil SE (2013) Common pitfalls in nanotechnology: lessons learned from NCI's nanotechnology characterization laboratory. Integr Biol 5(1):66–73
76. Cho EJ, Holback H, Liu KC, Abouelmagd SA, Park J, Yeo Y (2013) Nanoparticle characterization: state of the art, challenges, and emerging technologies. Mol Pharm 10(6):2093–2110
77. Brar SK, Verma M (2011) Measurement of nanoparticles by light-scattering techniques. TrAC Trends Anal Chem 30(1):4–17
78. Friedrich H, Frederik PM, de With G, Sommerdijk NAJM (2010) Imaging of self-assembled structures: interpretation of TEM and Cryo-TEM images. Angew Chem Int Ed 49 (43):7850–7858
79. Dobrovolskaia MA, McNeil SE (2007) Immunological properties of engineered nanomaterials. Nat Nanotechnol 2(8):469–478
80. Li J, Chang X, Chen X, Gu Z, Zhao F, Chai Z, Zhao Y (2014) Toxicity of inorganic nanomaterials in biomedical imaging. Biotechnol Adv 32:727–743
81. Fischer HC, Chan WCW (2007) Nanotoxicity: the growing need for in vivo study. Curr Opin Biotechnol 18(6):565–571
82. Aggarwal P, Hall JB, McLeland CB, Dobrovolskaia MA, McNeil SE (2009) Nanoparticle interaction with plasma proteins as it relates to particle biodistribution, biocompatibility and therapeutic efficacy. Adv Drug Deliv Rev 61(6):428–437
83. Burgess P, Hutt PB, Farokhzad OC, Langer R, Minick S, Zale S (2010) On firm ground: IP protection of therapeutic nanoparticles. Nat Biotechnol 28(12):1267

Top Med Chem (2016) 20: 75–104
DOI: 10.1007/7355_2015_91

Published online: 26 August 2015

Multimodal Imaging and Theranostic Application of Disease-Directed Agents

Joseph Caffarini, Nathan Kelleher, Christian C. Konopka, Madeline Mazurek, Anuradha Nandyala, Dwani Patel, Stephanie Slania, Sheryl Wang, Ravi Chandra Yada, and Dipanjan Pan

Abstract Contrast agents have long helped researchers and physicians alike delineate boundaries, but new diagnostic information is always sought after. A new field of molecularly targeted CT agents hopes to fill this void and supply physicians with prognostic information to find better treatments for patients. Borrowing from drug delivery and design, nanoparticles and similar platforms are being explored to help visualize complex biologic processes with never before seen resolution and fidelity. We discuss the development of this field and feasibility of translating some of these prospects to the clinic. Advances in chemistry, molecular biology, and engineering have molded this field emphasizing the early detection and treatment of diseases at the molecular and cellular level. Myriads of nanomedicine platforms have been proposed and developed and tested in laboratories and in preclinical models. However, very few have been translated to clinical trials. It is therefore a critical issue to recognize the factors affecting their eventual application in human. Better understanding of biological and biophysical obstacles encountered by these agents is necessary. Toward this aim, we critically review our present understanding of the biological obstacles encountered by the nano-agents, which we hope will motivate more studies to tune these technologies for future translational and clinical applications.

J. Caffarini, N. Kelleher, C.C. Konopka, M. Mazurek, A. Nandyala, D. Patel, S. Slania, S. Wang, and R.C. Yada
College of Engineering, Department of Bioengineering, University of Illinois at Urbana Champaign, Urbana, IL, USA

D. Pan (✉)
Department of Bioengineering, University of Illinois at Urbana Champaign, Urbana, IL, USA

Beckman Institute for Science and Technology, Urbana, IL, USA

Department of Materials Science and Engineering, University of Illinois at Urbana Champaign, Urbana, IL, USA

Carle Foundation Hospital, Urbana, IL, USA
e-mail: dipanjan@illinois.edu

Keywords Molecular imaging, Multimodal imaging, Theranostics, Therapy

Contents

Abbreviations

BBB	Blood-brain barrier
BLM	Bilayer lipid membrane
DCS	Differential centrifugal sedimentation
ECM	Extracellular matrix
ENM	Engineered nanomaterial
EPR	Enhanced permeability retention
HIFU	High-intensity focused ultrasound
i.v.	Intravenous
k_{off}	Dissociation rate constant
MPS	Mononuclear phagocyte system
MS	Mass spectrometry
NIPAM	*N*-Isopropylacrylamide
NLS	Nuclear localization signal
NMR	Nuclear magnetic resonance
NP	Nanoparticle(s)
PC	Protein corona
PEG	Polyethylene glycol
PS	Polystyrene
QCM	Quartz-crystal microbalance
RES	Reticuloendothelial system
SWCNT	Single-walled carbon nanotubes
Tf	Transferrin
TfR	Transferrin receptor

1 Introduction

The potential role of functional nanometer-sized agents in personalized medicine is undeniable [1–3]. Molecular imaging is demarcated as a noninvasive technique to observe cellular and subcellular events at a very early stage [4]. For the past

2 decades, the field has gained remarkable strength with high potential for clinical translation. This multidisciplinary area of research merges the major advancement in the areas of chemistry, molecular biology, genetics, and engineering to create unique opportunities to drive clinical imaging for early, sensitive detection, diagnosis, and treatment of a disease at the molecular and cellular level with unparalleled specificity [5, 6]. The potential of nanoparticles for both detection and drug delivery has been well documented [7–9]. Major advancement has been made toward the development of defined nanostructure for performing dual function, i.e., imaging and therapy (theranostics) [10–15].

2 Molecularly Targeted Agents for Imaging and Therapeutic Application

This chapter illustrates a presentation of the advancement related to the field theranostics, with major emphasis on computed tomographic (CT) and ultrasound-based imaging and therapeutics. We discuss the concept and introduce a few seminal works in these areas to judge the critical progress for basic, translational, and clinical applications.

2.1 Molecularly Targeted CT Agents: Feasibility, Prospects, and Path Forward

Computed tomography (CT) is a commonly used imaging technique because of its wide availability, efficiency, and cost [16]. In essence, CT generates a three-dimensional image from slices of X-ray images. Scanners rotate around the body, passing a thin X-ray beam through the patient to detectors on the other side. Like a normal X-ray, the waves pass through the body and lose energy. This decrease in energy is called attenuation, which varies based on the material through which it passes (bone, fat, etc.), resulting in differing colors in the image. More attenuation (i.e., fewer X-rays were allowed to pass through) creates white areas, whereas less attenuation (i.e., transparent materials) creates black images because the X-rays had enough energy to strike and change the silver halide film. While thickness and density play a role in attenuation, it depends more on the amounts of heavy (i.e., high-atomic-mass) metals in the tissue or material being imaged. By adding the slices of X-ray images together, contrast and resolution are improved [17]. The utility and pervasiveness of X-ray computed tomography (CT) has grown dramatically in the past 2 decades. Studies have shown that the use of CT has increased from 52 scans per 1,000 patients to 149 scans in the past 15 years alone [18]. The expanded use can be chalked up to several factors including advances in CT technology, reduced exposure, and expanded applications. CT provides an efficient platform to collect highly relevant physiological data with little to no invasiveness. Furthermore, the development of novel contrast agents has advanced CT into a high-fidelity, clear-cut contender in diagnostic imaging. We will explore some of the expanded applications of CT, specifically in regard to the use of new molecularly targeted CT agents.

In order to further enhance contrast, additional media can be added to the body. Negative contrast generally consists of air or other gases and will appear black on film, and positive contrast is the addition of a high-atomic-number material to appear white on film [19]. Common examples of positive contrast materials include barium, iodines, bismuth, and gadolinium, and it can be used on its own or in tandem with negative contrast media can be used to enhance a medical image [17]. Several qualities must be taken into account when judging the quality of a traditional contrast agent. Characteristics of the metal itself, such as atomic number and k-edge energy, are important. K-edge energy describes the binding of electron in the k shell around an atom. If the energy of an interacting photon, which occurs in X-ray imaging, is just above this binding energy, absorption and attenuation are increased and it is considered to be "well-placed" k-edge energy [20]. The material's interactions with the body are also taken into account. To be viable, agents must be biocompatible, have well-understood effects in the body, and be specifically sized. If too small, the renal system will break down or clear the material from the body too quickly, but if too large, it won't be cleared at all and could potentially be toxic to renal and other tissues [21]. A typical patient requiring a CT with contrast will have an iodinated compound as the contrast agent. Iodine's high attenuation helps physicians visualize complex and small regions to detect abnormalities. With a relatively short circulation time in the clinic, iodinated contrast agents are significantly restricted in their applications [20]. However, researchers are actively pursuing new forms of contrast agents aimed to address these issues and provide even greater prognostic information. Looking laterally at the drug delivery field, nanoparticles have been a promising vehicle for targeted treatment of illness, so active research is underway in using the nanoparticle platform to create novel contrast agents. Researchers hope to use the specificity and sensitivity of this platform to look at the molecular scale when investigating disease. There is an almost infinite array of possibilities and combinations of nanoparticle strategies for contrast agents due to the multitude of accessible polymer systems, targeting mechanisms, and conjugation strategies [20]. While other imaging modalities like MRI and ultrasound have several clinically approved molecular contrast agents, CT agents are still in the early stages of development. As with drug delivery, the use of nanoparticles raises worries of biocompatibility, clearance, cost, and long-term stability to name a few. The next generation of CT includes the idea of actively targeting contrast agents. Through the use of antibodies, ligands, or similar, the contrast media could be tuned to selectively accumulate on cells and tissues. The field shows promise, with studies involving gold, bismuth, and ytterbium nanoparticles which successfully label cancer targets [20]. By increasing contrast on a molecular level for this convenient and cost-efficient imaging modality, cancer therapy, including early detection and more accurate staging, can be significantly improved [16].

Molecular imaging has gained acclaim for its ability to allow researchers to look closely at microprocesses and understand the dynamics and factors influencing a cellular process [22]. With molecular imaging, we have an unprecedented ability to look at disease progression and the heterogeneity of manifestation. Diseases can affect individuals differently, and the differential development of that disease can directly affect the type of treatment that is required. With these tools, researchers have postulated the emergence of personalized medicine – a method to provide treatments specific to a patient. A "one treatment fits all" idea has long been noted as a limitation in current treatments of disease and cancer [5, 23–25]. Ongoing research in barium, bismuth, and binary contrast agents has yielded promising results, spurring continued investment in the field [20]. With continued development of molecularly targeted CT contrast agents, researchers are bringing a powerful weapon to a physician's diagnostic arsenal. Here we review the feasibility and prospects of molecularly targeted contrast agents and see what lies ahead for the field (Fig. 1).

2.1.1 Feasibility

Molecularly targeted CT contrast agents have the potential to make a significant impact on healthcare and diagnostics; however, there are a few barriers that are preventing their quick translation to the clinic. The current standard of care relies on nontargeted agents that are administered in relatively high doses and exhibit short in vivo circulation. Research in this field faces the challenge of creating cost-effective agents that are biocompatible, have high contrast efficacy, display long in vivo circulation times, and have long-term colloidal stability in physiologically relevant environments [5, 24, 25].

Biocompatibility is a major concern when considering the clinical translation of these agents. Iodinated nanoparticle agents pose a risk due to the high levels of iodine delivered to patients, which has the potential to potentiate radiation damage [26]. Gold nanoparticle agents provide a safer option for patients since gold is inert and considered nontoxic in vivo [27]. Compared to traditional CT agents, nanoparticle-based agents display longer in vivo circulation times, making them more feasible for a wider range of CT studies. By focusing on molecular targets, these targeted agents can be used in smaller doses than traditional agents, making them safer for patients. Another setback with nanoparticle-based agents is the efficient delivery of contrast to the patient. Due to the nature surface covalent conjugation, iodine nanoparticles are unable to carry a large load of iodine on its surface [5, 24, 25]. In contrast, gold nanoparticle agents have the ability to load large amounts of gold on the surface, decreasing the concentration of nanoparticles needed to be delivered but still posing a major obstacle in clinical translation due to the cost of gold [5, 24, 25].

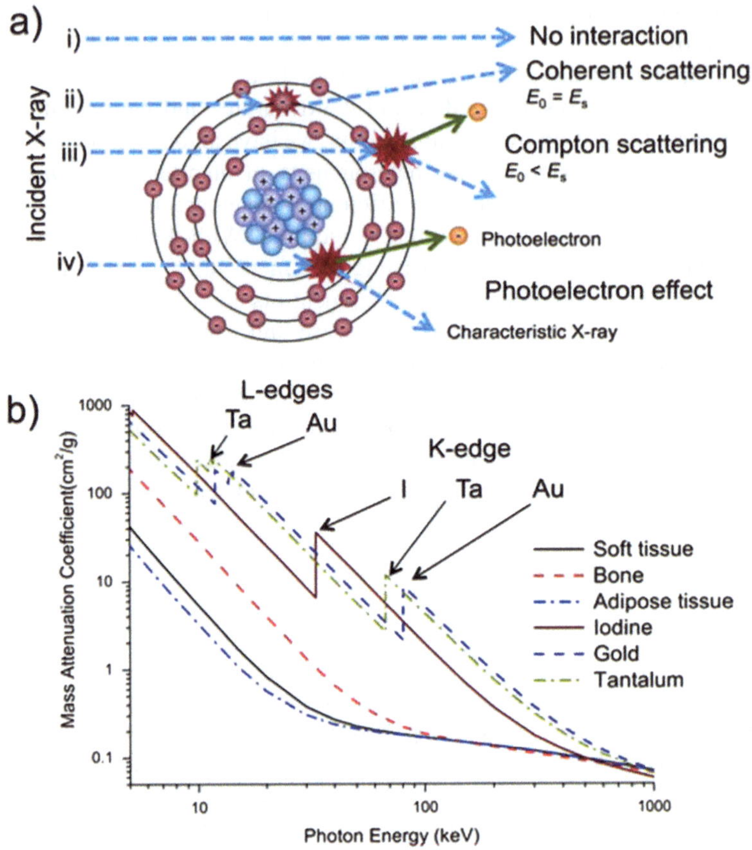

Fig. 1 (**a**) Various interaction with X-ray with matters, including transmission, coherent scattering, compton scattering, and characteristic X-ray radiation. (**b**) Mass attenuation of several materials. Reproduced with permission from [21]

2.1.2 Current Strategies

Although CT provides excellent spatial resolution, depth penetration, and relatively rapid image acquisition, its present applicability to molecular imaging is limited [28]. As described above, however, integration of recent advances in CT technology, in nanomedicine, and in spatiotemporally controlled agent delivery may greatly enhance the potential of CT for molecular imaging toward earlier, more personalized disease detection. Fusion imaging is an increasingly common strategy to overcome the molecular limitations of CT imaging in the clinic. Integrated positron emission tomography/CT (PET/CT) imaging capitalizes on the anatomical and functional information provided by individual modalities, enhancing early diagnosis and immediate observation of therapeutic responses particularly in the heart, the brain, and cancer cells [28]. PET takes advantage of the radioactive decay

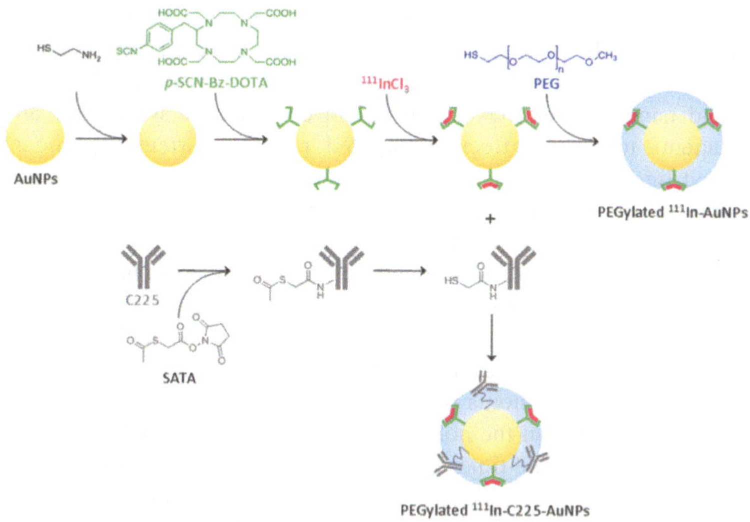

Fig. 2 Process of tagging AuNPs with various chemical agents [31]. A variety of chemical groups can be added to specify different targets and allow for different chemical properties

of a tracer on a biologically active molecule, most commonly fluorodeoxyglucose (FDG) for cancer detection. FDG accumulates in tissues with high glucose uptake, including the brain, liver, and most cancer cells. Co-registration with CT allows anatomic identification of abnormal glucose uptake, presenting molecular indications of cancer [28]. PET/CT has become a powerful clinical tool for early disease diagnosis; however, it is limited by high instrumentation costs and potentially severe radiation exposure.

Enhancing the sensitivity and contrast of CT imaging will expand the capabilities of CT beyond structural imaging to include functional and molecular imaging. As one of the leading diagnostic technologies in terms of cost, availability, and efficiency, adding molecular imaging capabilities to conventional CT will have far-reaching and significant implications for earlier and more sensitive disease detection [29]. Recently, high-atomic-number metal-based nanoprobes, such as gold nanoparticles, have been developed as blood pool contrast agents and have demonstrated strong X-ray attenuation in preclinical animal models, resulting in significant contrast enhancement [29]. Tuning the size and surface functionalization of these probes to enable passive and active targeting further enhances the specificity of CT. A paradigm shift in clinical disease management from post hoc symptom-based treatments toward preventative and personalized medicine necessitates the development of molecular imaging capabilities for conventional modalities [30]. Recent advances in CT technology and design of nanoparticle-based contrast agents position CT as a promising tool for accessible and efficient clinical molecular imaging (Fig. 2).

Gold nanoparticles have many properties that make them useful for molecular therapy. For instance, they are nontoxic and biocompatible and absorb light in the

NIR and X-ray spectrums [32]. The optical properties allow the particles to be used in therapy, where NIR light can heat the nanoparticles to kill the tumor or deliver medication, and the X-ray absorbance creates imaging contrast [31]. Such properties of gold nanoparticles make it useful in combination with other atoms and contrast agents to create effective targeted probes. Current experiments have incorporated ligand-targeted gold nanoparticles (AuNPs) to create specific tumor imaging probes for cancer diagnosis. Functional groups are added to AuNPs by exploiting the strong and specific interaction between gold and sulfur [31]. Because of the selective formation of sulfur bonds, a large number of chemical tags can be added to AuNPs, including hormones, antibodies, other chemical contrast agents, or other receptor-specific peptides. For example, Kao et al. created PEGylated AuNPs tagged with antibodies targeted for epidermal growth factor receptor (EGFR) to target malignant lung carcinoma tumors in mice [33]. The targeted particle uptake was 14.9 times higher than the nontargeted control in high EGFR-expressed lung carcinoma cells and was 3.8 times higher in lower EGFR-expressing breast adenocarcinoma cells [33]. The particles also stayed in the tumors for longer than nontargeted agents and were taken into the cells via antibody-mediated endocytosis [33]. The ability of AuNPs to be specifically targeted to cancer cells and be absorbed through endocytosis means they can be used both in diagnostics and in chemotherapy. A specific example cited by Kao et al. is the tagging of AuNPs with cetuximab to kill mouse pancreatic tumor cells in the presence of radio frequency radiation [33]. AuNPs can also be tagged with other chemical groups for multimodal imaging analysis. Radiolabels can be attached to the probes to allow for SPECT overlayed CT images and allow for specific targeting of radiotherapy agents [33]. Gold nanoparticles prove to have diverse applications in medicine and will eventually be powerful tools in cancer treatments (Fig. 3).

Bismuth nanoparticles are another attractive option for therapy and imaging due to their lower cost. Bismuth salts have been used as imaging agents for X-ray since the late 1800s. However, these salts quickly became toxic at high concentrations. Polyvinylpyrrolidone (PVP)-coated bismuth particles, however, are safe to use and can be used as CT contrast agents even at higher concentrations [34]. These particles have high X-ray absorption as well as long circulation times. The PVP coating is essential for these nanoparticles. The uncoated particles were seen to aggregate at physiological pH, have a low final concentration that was not suitable for X-ray contrast, and have a very low circulation time in the body. Due to the high X-ray absorption, lower concentrations of the bismuth nanoparticles can be used than conventional iodine methods as shown in Fig. 4. When injected into bald mice, the PVP-coated bismuth nanoparticles had a half-life of 140 min, which is also an enhancement upon iodine particles. In terms of clinical applications, bismuth nanoparticles can be used to image vasculature, tumor angiogenesis, as well as multivalent targeted imaging agents. Rabin et al. demonstrated the use of bismuth nanoparticles in the detection of hepatic metastases. Mice were intravenously injected with the nanoparticles and imaged. The results showed that the particles accumulated in the liver due to their uptake by phagocytes and hepatocytes. They also tested whether the particles could be used for lymphatic cancer staging. After

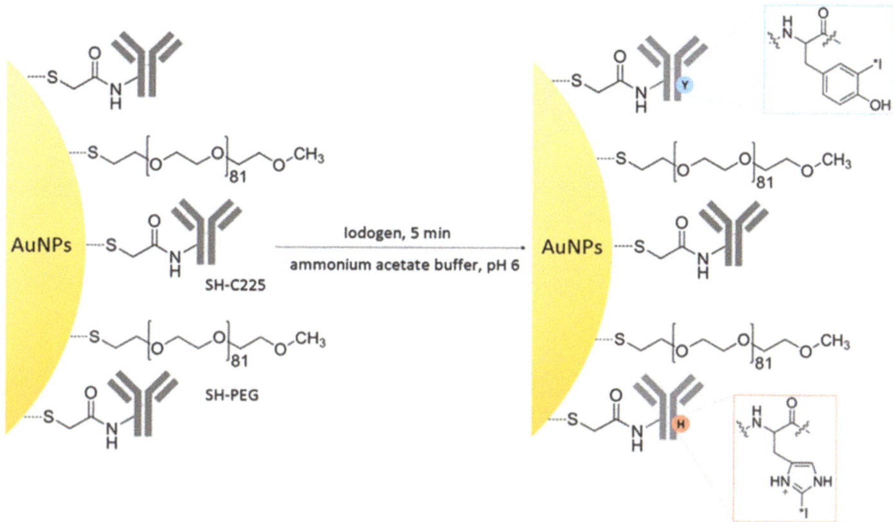

Fig. 3 Tagging AuNPs with radioactive iodine [31]. Two different isotopes were used, iodine 131 and iodine 123; the specific atoms are marked with an asterisk in figure. Iodine 131 exhibits a beta emission of 0.606 MeV, allowing it to be used in nuclear treatments. Iodine 123 gives off less radiation than iodine 131 and can be retained in cells and used as a radiolabel in CT/SPECT imaging

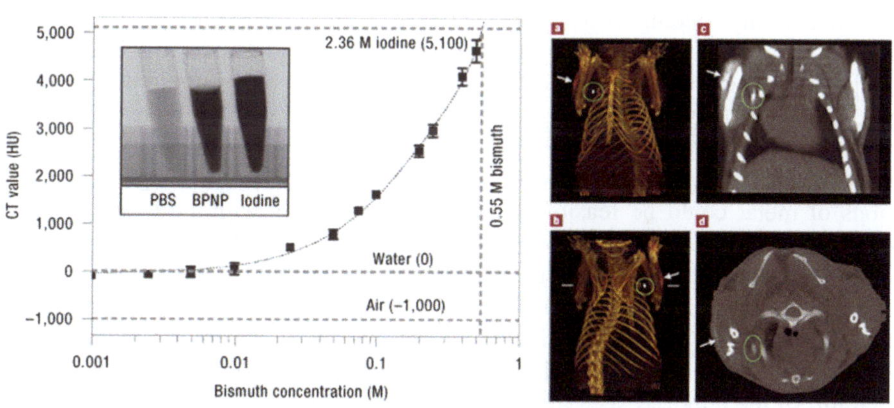

Fig. 4 (*left*) Calibration curve of X-ray attenuation; (*right*) imaging after lymph node administration of bismuth nanoparticle

administration to the mice, lymph nodes were clearly highlighted by the particles as can be seen in Fig. 4 (right).

Moreover, Pan et al. further demonstrated the clinical value of bismuth through research on NanoK-enhanced spectral CT molecular imaging, which can be utilized to achieve one of the major aims of cardiology, the detection and quantification of

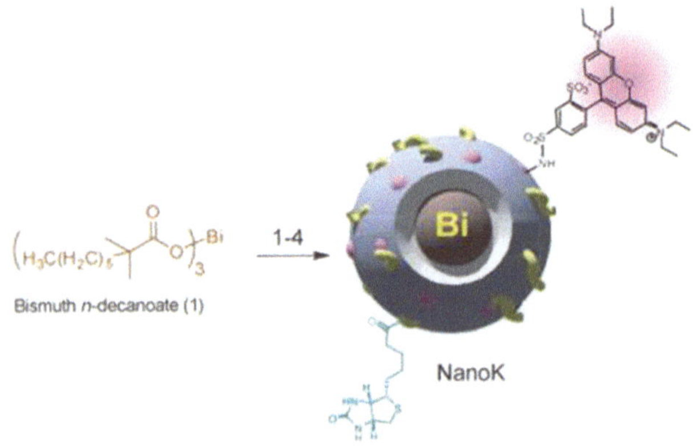

Fig. 5 NanoK synthesis. Preparation of bismuth-enriched K-edge nanocolloid (NanoK (Bi)): (1) suspension of bismuth *n*-decanoate (1) in sorbitan sesquioleate, vigorously vortex and mixing, filter using cotton bed, vortex; (2) preparation of phospholipid thin film; (3) resuspension of the thin film in water (0.2 μM); (4) microfluidization at 4°C, 20,000 psi, 4 min, dialysis (cellulosic membrane, MWCO 20 K)

ruptured plaque in coronary arteries [35]. Via active targeting mechanisms, NanoK can deliver high concentrations of metal to a specific site, with tunable payloads between 40 and 70% v/v; see Fig. 5. Using spectral CT, the specific targeting of NanoK, stemming from receptor-ligand interactions, to fibrin on carotid artery endarterectomy specimens was observed in vivo, thus achieving the spectral contrast necessary for clinical applications. In addition to conveying information graphically, spectral CT also provides quantitative data (concentration of bismuth per voxel) and may therefore provide clinical value by identifying pressing health risks. Furthermore, whole-body bio-elimination studies were conducted on adult male mice after intravenous injections, and they verified that the high concentrations of metal could be feasibly cleared from the body within 14 days without causing damage to the liver or kidneys. In another study, Kinsella et al. demonstrated the use of targeted bismuth nanoparticles to image breast cancer tumors. They utilized a nine-peptide chain (LyP-1) to target 4T1 cancer cells in mice. This cyclic peptide specifically targets the gC1q receptor p32 protein in cancer cells. This targeting of the bismuth particles allows clearer imaging of the margins of the tumor as seen in Fig. 6.

Overall, the clinical value gained from bismuth nanoparticles is significant, and as adjacent technologies concurrently develop, the medical impact will increase. Applications have already been developed to treat a range of diseases (from cardiovascular disease to cancer), and future innovations offer promise.

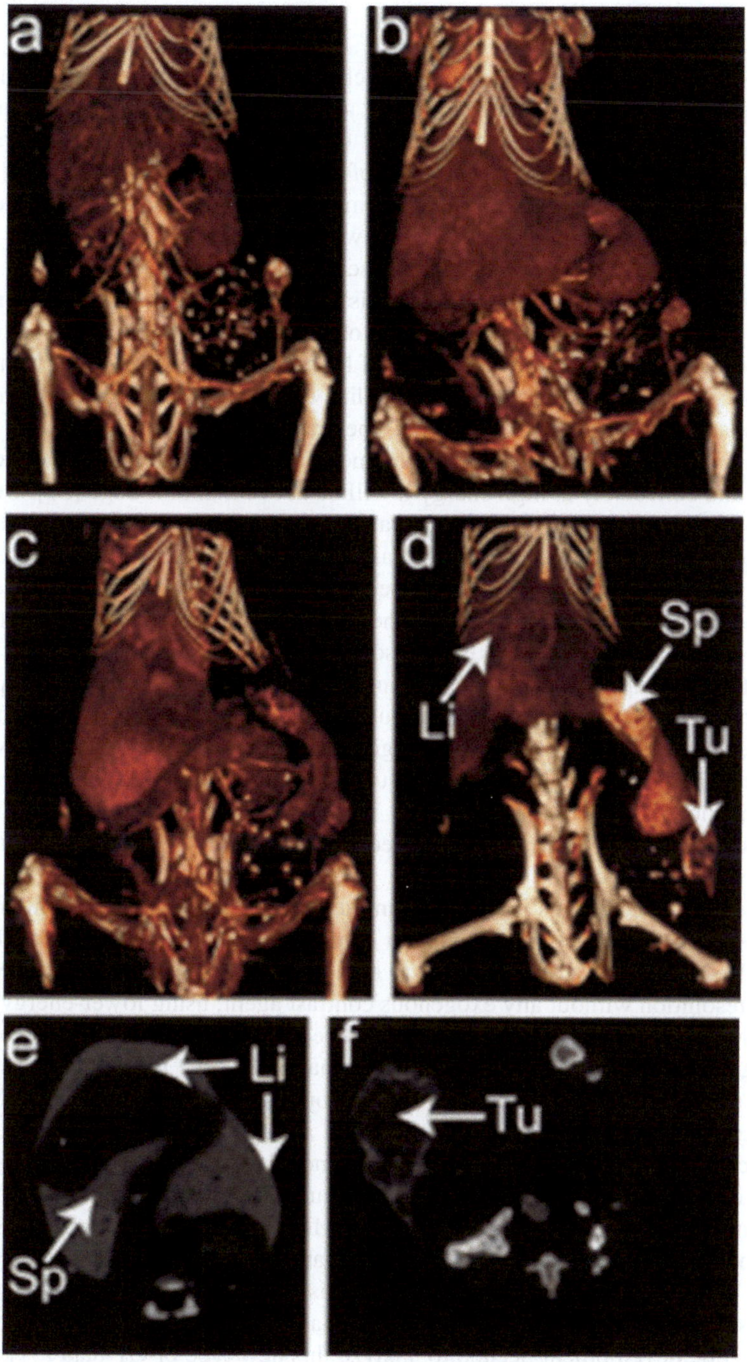

Fig. 6 microCT image of cancerous mouse using LyP-1-targeted bismuth nanoparticles

2.1.3 Future Directions

CT has been used to great effect in both clinical and preclinical worlds due to its relatively low cost, high spatial resolution, and quick scanning time. Unfortunately, CT imaging has previously been limited due to its low soft tissue contrast and consequent need for high concentration of exogenous contrast agents [36]. However, with the recent surge of molecular imaging techniques and new CT technology, the field of imaging experiences a newfound interest in expanding the uses of targeted contrast agent CT scans and phase-contrast imaging [36].

One such technological development is spectral CT. Spectral CT is a method which exploits the unique attenuation factors each element might have at different X-ray energies, by using a combination of high- and low-energy X-ray beams. For example, under normal circumstances, iodine contrast and bone calcium are indistinguishable in a CT scan, but using spectral CT, these two materials can be distinguished because at low energy the mean energy will be below the k-edge of iodine and conversely at high energy it will be above that of k-edge [37]. One such example would be where a group used a dual energy CT (DE-CT) with iodine and gold nanoparticles as exogenous contrast agents to quantitatively measure tumor blood volume and vascularization, as seen in Fig. 7 [37]. Cancer imaging such as this and countless other applications where targeted or nontargeted contrast agents can be used for enhanced soft tissue resolution are what makes spectral CT such a powerful technique and of such an interest for future research developments. Unfortunately spectral CT alone without exogenous contrast agents does not provide beneficially higher-resolution imaging; therefore, researchers have developed another technique to enhance soft tissue resolution without exogenous contrast. This method is called phase-contrast imaging. In phase-contrast imaging, the X-rays detected by CT are viewed as electromagnetic waves instead of as particles, as is traditionally done by CT.

This difference results in a change in the way scientist will represent an X-ray beam's index of refraction, attenuation coefficient, and phase change coefficient. With these differences, it is possible to obtain images with a higher degree of soft tissue resolution without any exogenous contrast agent, using lower-energy X-rays [36]. So far, this method has been proved to be valid, but the technology remains in its beginning stages. One such experiment that proves the feasibility of this technique compares different types of phase-contrast gating methods to MRI and classical attenuation-based CT imaging in its ability to visualize pancreatic ductal adenocarcinoma (PDAC) tumors in mouse models. As can be seen in Fig. 8, the phase-contrast images provide robust soft tissue differentiation compared to that of conventional CT and comparable to that of MRI. Most importantly phase-contrast imaging is able to identify both qualitatively and quantitatively the presence of the tumor in soft tissue [38]. The method so far is still limited greatly due to it taking large scan time (up to 10 h), fabrication challenges to scale fields-of-view, and algorithmic development needed to analyze this new type of CT data [37].

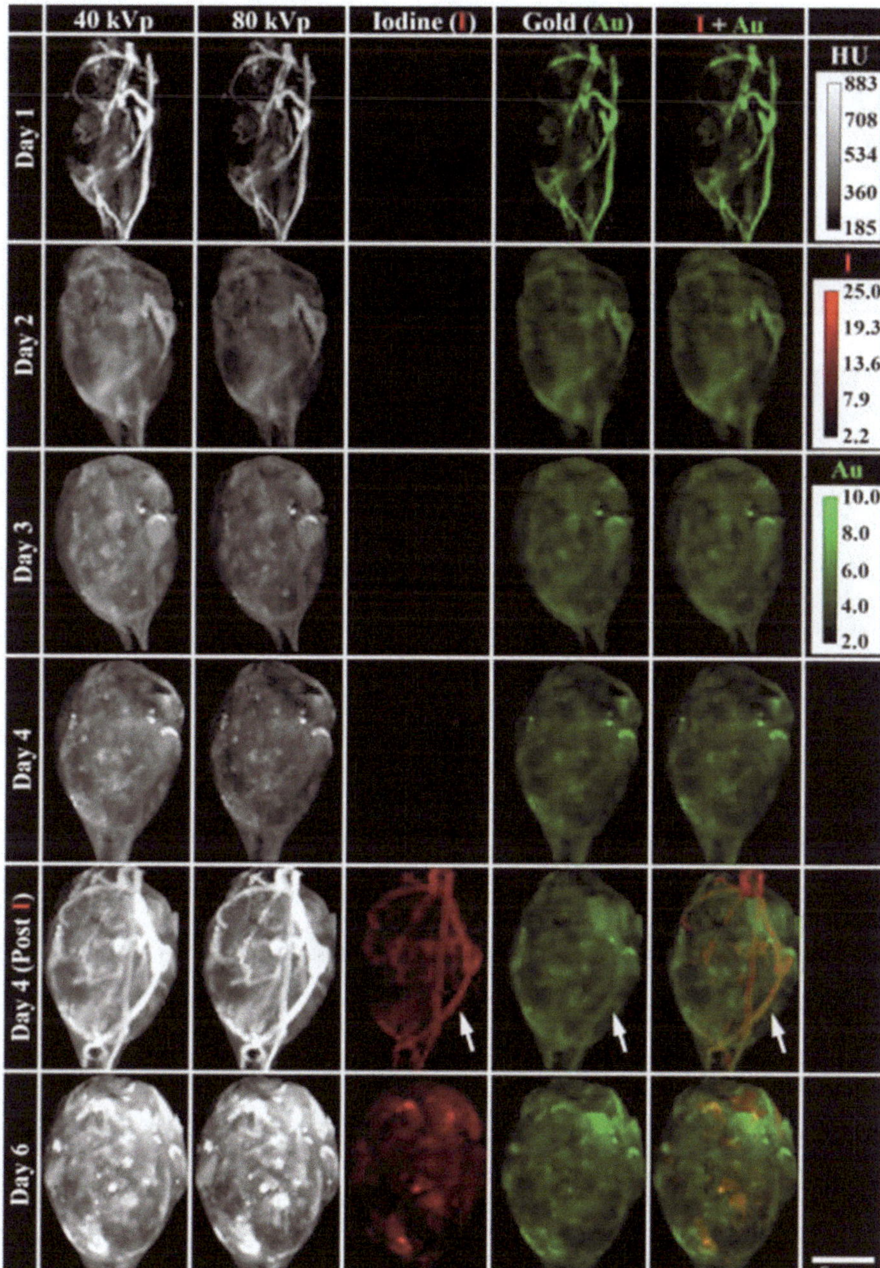

Fig. 7 In vivo image of a primary soft tissue sarcoma using DE microCT: day 1 (gold nanoparticle injection), day 2, day 3, day 4 (pre-liposomal iodine injection), day 4 (post-liposomal iodine injection), and day 6 demonstrate the simultaneous decomposition of I and Au in vivo

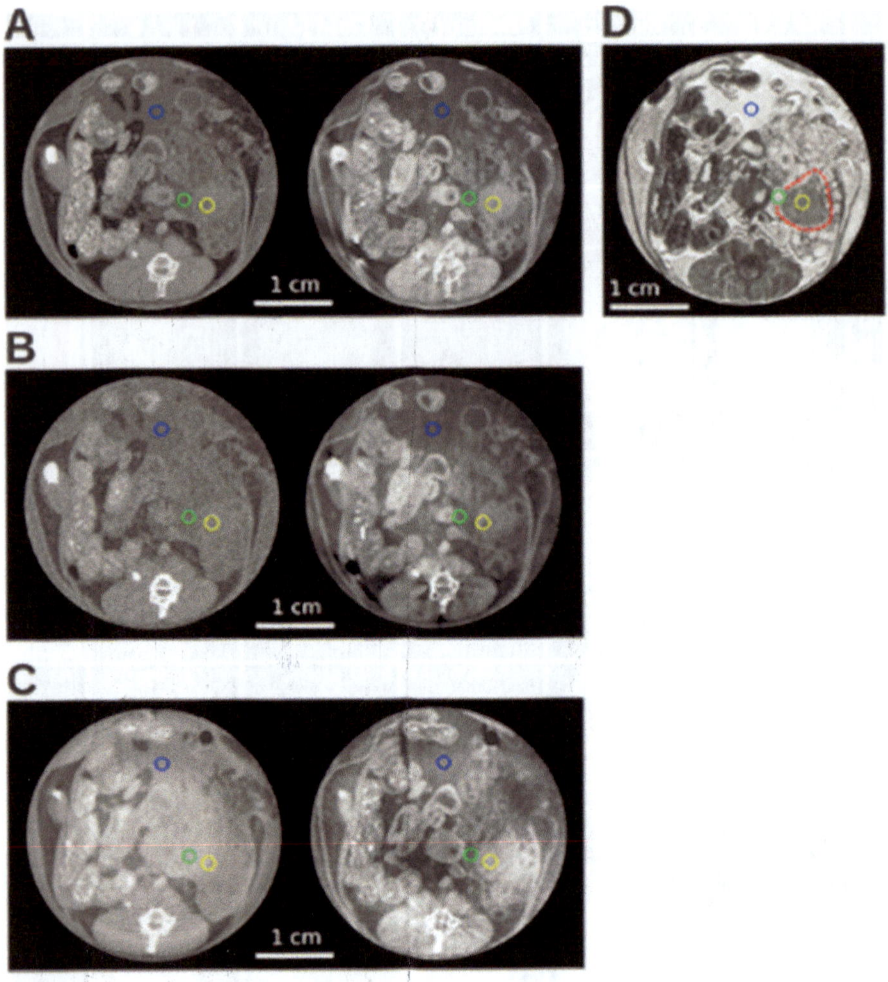

Fig. 8 Analysis of tumor visibility as identified by *red line* in MRI image (**d**) with (**a–c**) traditional attenuation-based CT (*Left*) phase contrast (*Right*) and multiple sources: (**a**) high-resolution synchrotron source. (**b**) Low-dose synchotron source (**c**). Tube source. (**d**) MRI

CT technology is on the cusp of great translational advances in the medical field. Not only have the current methods used in CT been greatly enhanced through nanotechnology and molecular imaging strategies, but the future of CT imaging using spectral CT and phase-contrast CT is very promising. With these newfound techniques and technologies, the medical field will certainly see novel use for CT in almost every field of medicine. Growing knowledge of cellular biology, pathophysiology, and disease progression has demanded the development of high-resolution, sensitive, and reliable clinical and preclinical biomedical imaging. CT is a popular diagnostic tool high spatial resolution. However, CT alone is predominantly used to image hard tissue. With the development of molecularly CT agents, one can visualize the anatomy and functionality with far greater detail and accuracy. Molecularly targeted CT agents will expand the research, diagnostic, and treatment capabilities.

2.2 *Ultrasound-Mediated Therapy: A Review of Preclinical and Clinical Applications*

Ultrasound is one of the most heavily utilized imaging modalities in the clinic today. The lack of ionizing radiation and its advancing resolution capabilities have kept the modality thriving since its conception in the late 1930s. The scope of ultrasound has been explored and expanded in the decades since, and now it is being developed into a powerful tool for therapeutic applications. With promising pre-clinical outcomes thus far with new ultrasound-mediated theranostics, the field has grown to advance ultrasound from its diagnostic roots to therapeutic intervention. Here we review the preclinical and clinical works of ultrasound-mediated therapy, focused on current strategies and where the field is heading.

Ultrasound (US) as a biomedical tool has diagnostic and therapeutic applications. US uses sound waves at frequencies above the range of human hearing, 20Hz–20kHz (Fig. 9). In medical imaging, US offers tomographic and real-time views of anatomy at a relatively lower cost than magnetic resonance imaging (MRI) or computed tomography (CT). At frequencies from 800 kHz to 2 MHz, therapeutic US utilizes high-frequency sound waves in order to stimulate tissue. Therapeutic US offers many uses that are beneficial to healthcare and medicine. US can assist photodynamic therapy to treat cancer, break down kidney stones to make them easier to pass, and can even help with liposuction. Low-intensity pulsed US has been shown in aid in the stimulation of bone regeneration. In addition, US also present theranostic benefits when applied with microbubbles. The controlled administration of high-frequency US can induce microbubble cavitation and release encapsulated contrast agents. Moreover, microbubbles can be targeted to specific tissue for controlled drug release. One major application of these theranostic microbubbles and focused US (FUS) techniques has been to penetrate the blood-brain barrier – a major hindrance in the diagnosis and treatment of neurodegenerative diseases and brain cancer.

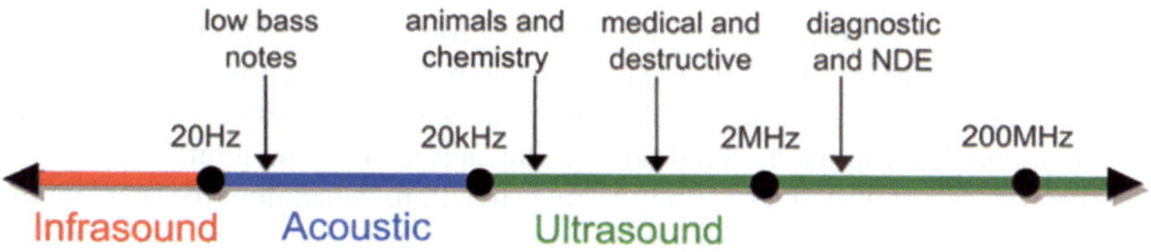

Fig. 9 Frequency spectrum illustrating US range

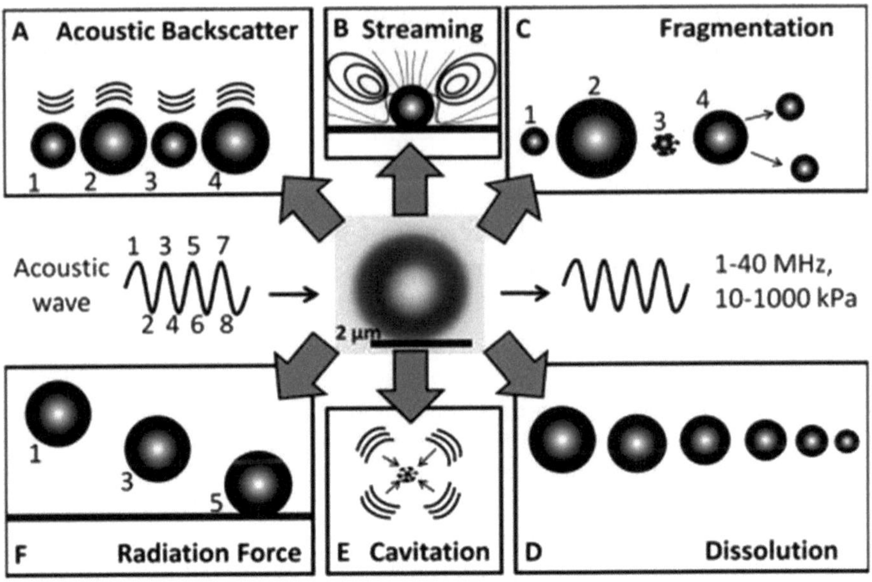

Fig. 10 Ultrasound-mediated effects on microbubbles for diagnostic and therapeutic applications [41]

The holy grail of the drug delivery field is a therapy that will be able to deliver a therapeutic cargo to an extremely specific site, tissue, or even cell with high efficacy while minimizing side effects like healthy tissue toxicity and cellular response [39]. Researchers have poured over this idea, spurring new fields, and have developed novel methods to address this goal. Now, researchers are looking into using ultrasound-directed therapy with known contrast agents as a viable option.

The most common and developed ultrasound contrast agent and future drug delivery vehicle is the microbubble. Microbubbles are gaseous containers with a hard coating which are responsive to ultrasound energy [40]. Upon intravenous injection, they can travel in the bloodstream and can be activated by directed ultrasound at a site of interest [39], the idea being that the microbubbles will reflect ultrasound waves in a unique manner, thus allowing you to distinguish the bubbles from the surrounding tissue [40]. This simple idea has long been used in the diagnostic paradigm; however, the discovery of novel ultrasound-mediated effects (Fig. 10) has dramatically expanded its uses [41]. As shown in Fig. 10b, streaming leads to local mixing of compounds and creates shear forces leading to enhanced intracellular and extravascular transport of macromolecules [41]. Fragmentation, dissolution, cavitation, and radiation force are all properties employed for enhanced imaging, but have vast potential to be tailored to assist drug delivery mechanisms.

Microbubbles can be tailored to specific applications in terms of delivery mechanism, site of delivery, and desired function. Changing the gas in the microbubble, coating, and loading all affect the function of the microbubble.

Combinations of these factors have been exhaustively tested for drug and gene delivery applications. We will look into some of the successful approaches developed thus far and their application in both the preclinical and clinical setting.

2.2.1 Clinical Relevance

Ultrasound imaging has become nearly ubiquitous in clinical settings, particularly in cardiology and obstetrics. Due to the low risk and relatively low cost of ultrasound technology, integration of current ultrasound systems with therapeutics represents a promising avenue to broadly and rapidly enable image-guided therapies and theranostics in the clinic [41].

Microbubbles are being extensively studied for gene, small molecule, and protein drug delivery and as therapy through mechanical destruction of bubbles by ultrasound [42]. Researchers have entrapped plasmid DNA in the polymer coating of microbubbles, allowing sustained release by diffusion through the matrix as well as rapid release by applying ultrasound to cyclically compress and expand the microbubbles [41]. These microbubbles encapsulate their cargo, often protecting nucleic acids from enzymes in the blood and extending the lifetime of gene therapies in the physiology. In addition to enhancing in vivo circulation times for gene therapies and drugs, microbubbles enable external spatiotemporal control of the release of these agents using ultrasound. Molecular and cellular targeting, such as activated leukocytes for inflammation, further localize microbubble-delivered therapies [42].

Despite the convenience and versatility of potential ultrasound-mediated therapies, several barriers to clinical translation exist. Biocompatibility of microbubbles is of paramount importance, as such, proteins, lipids, surfactants, and polymers have been explored as potential microbubble materials in the preclinical and clinical setting [42]. Microbubbles are additionally restricted to the bloodstream, limiting applications in bulk tissues, but making them ideal for cardiovascular targets. Cavitation and eventual mechanical destruction of microbubbles has been shown to cause capillaries to collapse and to cause cellular damage [42]. These effects may have both positive and negative implications for therapy, aiding tumor destruction, but increasing risk of unintended vessel and tissue damage in other cases.

2.2.2 Current Strategies

One of the major therapeutic applications of ultrasound has been in mediating drug delivery, which can be achieved using both thermal and mechanical mechanisms. Focused ultrasound has the ability to increase the temperature at a targeted region which has been shown to increase vascular permeability and blood flow at that site [43]. This has important implications in cancer drug delivery, where delivery of drugs to tumor sites could be increased by increasing blood flow and permeability of the tumor's vasculature, making the treatment more effective. In addition, temperature-sensitive liposomes (TSLs) have been used in combination with focused ultrasound as a promising strategy for drug delivery. Liposomes act as a protective barrier to help facilitate the flow of the drug through the bloodstream, minimizing its clearance and nonspecific uptake. Once the TSLs arrive at an ultrasound-heated site, they quickly dissolve and release the drug from the liposome, delivering the drug at the site of interest [44]. Although increasing drug delivery via ultrasound-induced hyperthermia has proven to be a successful strategy, the effectiveness of this mechanism cannot be confirmed due to contradicting results [45].

Microbubbles and their use for drug delivery is a highly investigated mechanical mechanism of ultrasound therapy. This mechanism works by combining ultrasound with gas-encapsulated microbubbles to cause openings in a nearby cellular barrier such as the cell membrane or the blood-brain barrier, a process called sonoporation [46]. The disturbance in the surrounding barrier is temporary and is caused by the "popping" of the microbubble in response to the ultrasound waves. By disturbing these membranes, drugs can pass easily to previously inaccessible targets, opening the door to new treatments and therapies. To increase the efficiency of drug delivery, drugs are loaded on to the microbubbles before undergoing sonoporation. Depending on the shell of the microbubble (lipid, protein, or polymer based), different loading strategies are utilized to capitalize on the properties of the microbubble, as seen in Fig. 11. Typically, these drugs are either loaded onto the surface of the microbubble or encapsulated within the internal void. In addition, surface modifications, such as the addition of targeting ligands, allows microbubbles to target specific sites, thus increasing the amount of drug delivered at the site of interest and opening the door to numerous applications for this strategy [46].

US has various applications in the medical field. Its nonionizing, affordable, portable, and noninvasive nature makes US one of the most utilized imaging modalities, but it has since spread into the therapeutic sector of medicine as well [47]. There are two main methods by which US has effect: thermal and nonthermal. When the US energy is absorbed by the tissues, they are heated. This temperature shift is controlled by having longer durations of exposure with unfocused beams and is utilized most in physical therapy to produce enhanced healing. Nonthermal applications include ultrasonic cavitation, microstreaming, and gas body activation [48]. Microstreaming flow creates shear stress on the cells, leading to cell lysis. In vivo gas body activation generates intracellular microstreaming and thus acts by the same mechanism [49]. Cavitation is the formation of bubbles within tissues and body fluids which can be either stable or unstable. Unstable cavitation can enhance acoustic streaming, whereas unstable cavitation is the release of energy from collapsed bubbles (Watson). The creation of stable vs. unstable cavitation can be controlled depending on use. The formation of both is illustrated in Fig. 12. Therapeutic US varies the wave characteristics (e.g., amplitude, frequency, propagation length) to create specific biophysical effects. For example, increasing frequency or pulse length will increase heating and thus is utilized for thermal

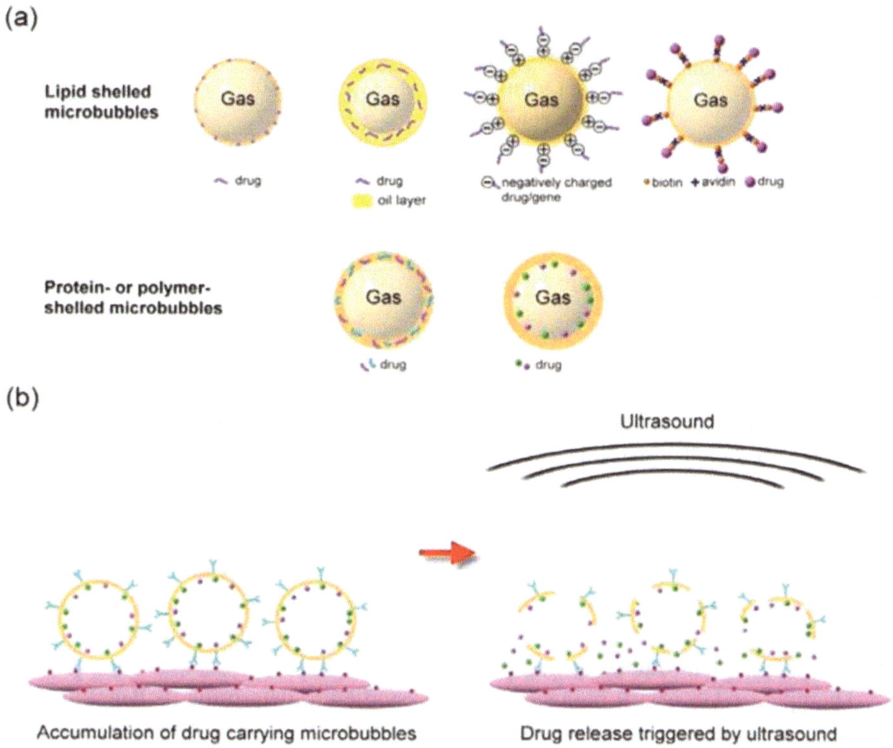

Fig. 11 Overview of drug loading and delivery methods of microbubbles [46]

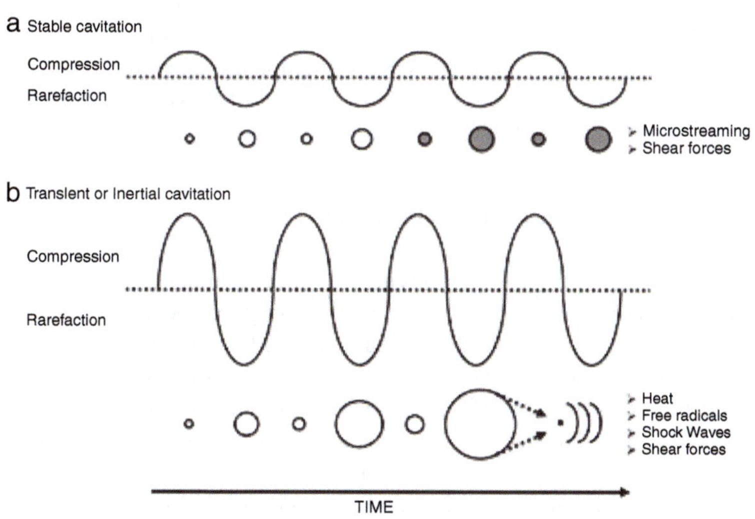

Fig. 12 The formation of stable and unstable cavitation and their therapeutic uses [50]

Table 1 The penetration of US waves as it varies with frequency and medium (Watson)

	1 MHz	3 MHz
Muscle (mm)	9.0	3.0
Fat (mm)	50.0	16.5
Tendon (mm)	6.2	2.0

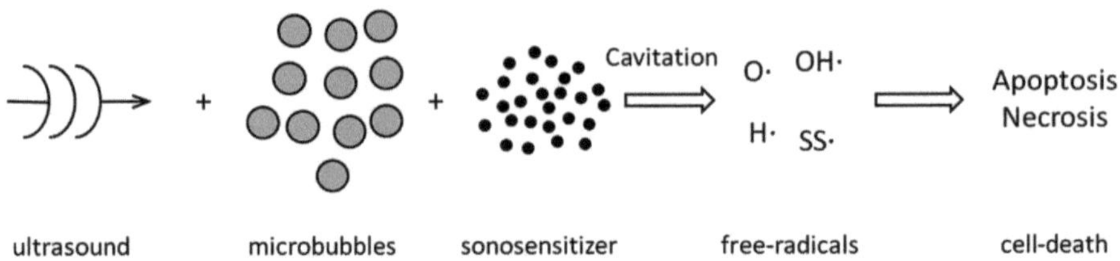

Fig. 13 Sonodynamic therapy mechanism. The generation of free radicals from microbubble cavitation leads to cell apoptosis and death [52].

applications. For cavitation, short, focused, high-intensity US pulses can break up soft tissue. The characteristics of therapeutic US also depend on the medium through which the waves are traveling [48]. Depending on the material (e.g., fat, muscle) and the desired depth of penetration, some frequencies will produce more desired effects than others, as seen in Table 1.

Therapeutic US is still being extensively studied to determine its benefit to risk ratio for patients with various conditions. Currently, there is no known cumulative dose, or effects from repeated exposure, defined for US therapy. Some side effects, including burns, vascular injury, and scarring, have been noted [48]. However, there are several studies questioning this method's effectiveness. One review demonstrates the inability of US to treat pain, musculoskeletal, and soft tissue conditions better than a placebo US [51]. The review is limited to those specific fields, and for many other functions, US for therapeutic purposes is both accepted and beneficial. In the following section, several of those widely utilized US applications will be covered in further depth.

2.2.3 Current Strategies and Applications

Sonodynamic Cancer Therapy: Sonodynamic cancer therapy uses low-intensity US pulses at diagnostic frequencies to create reactive oxygen species which kill rapidly dividing cancer cells. The method is based on photodynamic therapy, where light is used to generate free radicals. Sonodynamic therapy is more effective; during trials it hindered tumor growth by 77%, whereas photodynamic therapy only inhibited growth by 27% [52]. The use of US waves is more effective than light because US penetrates deeper into tissue to access tumors [52]. During therapy, microbubbles are administered with a sonosensitizer and are cavitated by low-intensity US waves. The cavitation causes the sonosensitizer to generate radicals which then react with cells and lead to apoptosis, as shown in Fig. 13. The therapy typically uses passive

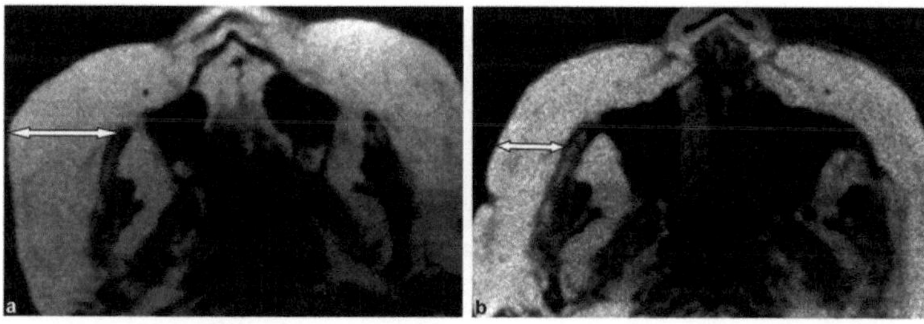

Fig. 14 (**a**) Preoperative and (**b**) postoperative MRIs of a patient undergoing siliconoma removal. The FUS loosen the surrounding softer tissue so the siliconoma can be removed [54]

targeting mechanisms of intravenously injected microbubble-sonosensitizer mixtures.

Shock Wave Lithotripsy: FUS waves can be used to break up kidney and ureter stones. The basic mechanism occurs when a strong US pulse hits a stone-water interface and generates longitudinal (P) and transverse (S) stress waves which propagate inside the stone [53]. When these waves reencounter the stone-water interface, they generate more P and S waves which constructively overlap to generate high levels of stress in the stone, causing it to break down [53]. Cavitation bubbles also form in the surrounding fluid from the negative pressure components of the incident acoustic wave [53]. This process helps in breaking down the surface of the stone. Clinical trials have proven this method to be safe and effective; however there are a few potential side effects: the microbubbles generated from cavitation can induce vasoconstriction and free radical formation, leading to local ischemia and tissue damage. Sound wave lithotripsy does not work for all stone compositions, such as calcium oxalate monohydrate. Moreover, the procedure is often paired with CT to identify the stone composition before treatment.

Ultrasound-Assisted Liposuction: FUS waves can also be used to break up fat tissue during liposuction and siliconomas resulting from cosmetic surgery complications as seen in Fig. 14 [54]. Frequencies greater than 16 kHz are used to cause cavitation specifically within fat tissue, causing a breakdown of cellular structure [55]. The fat cells are then suctioned out. The fat cell cavitation with US-assisted liposuction (UAL) decreases blood loss, operative time, bruising, and discomfort particularly in fibrotic areas like the chest and side [55]. However, UAL patients have an increased risk of burns and will need to have larger incisions [55].

Theranostic Applications of Microbubbles and Ultrasound: The combination of microbubbles (comprised of two structural components – an encapsulating shell and an inner gas core) and acoustics provides numerous diagnostic and therapeutic applications – including microbubble contrast agents used for clinical imaging and microbubble site-specific delivery systems that can be loaded with genes or drugs; see Fig. 15 [56]. Microbubbles significantly improve the signal-to-noise ratio of images, as they are compressible spheres that are easily distinguishable from surrounding tissue due to their unique nonlinear oscillations in response to an

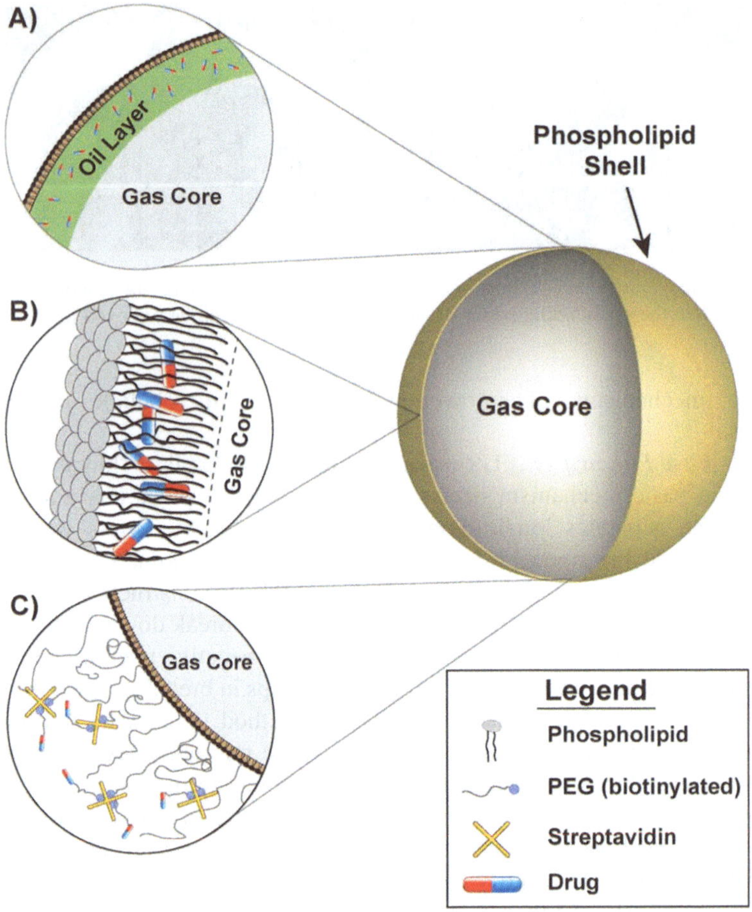

Fig. 15 Schematic of commonly used drug attachment strategies in microbubble-mediated drug delivery. (**a**) Drugs can be dissolved in a secondary oil layer using a multilayer microbubble construction. (**b**) Therapeutic agents can be seeded within the thin encapsulating shell. (**c**) Nanoparticles or other therapeutics can be attached to the outside of the shell, such as tethered to PEG chains [56]

applied acoustic field. Moreover, microbubbles can be used to enhance therapeutic delivery, as their inertial cavitation and destruction produce a strong mechanical stress that manipulates the permeability of vasculature and cell membranes [57]. Ongoing studies aim to capitalize on these properties to design novel microbubble vectors for in vivo and targeted gene delivery [58]. In addition, microbubbles can actively target specific biomarkers via the incorporation of ligands as well as selectively ablate tissue through the enhanced conversion of acoustic to thermal energy [56]. To further illustrate the broad clinical applicability of microbubbles, they are capable of enhancing contrast in blood perfusion imaging

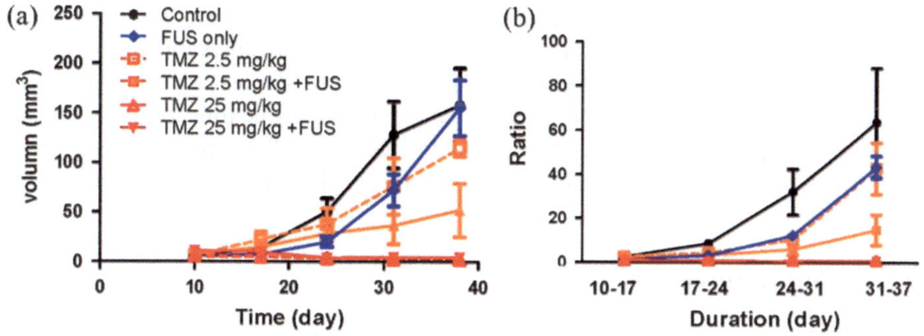

Fig. 16 (**a**) Tumor progression (in volume, mm^3) from days 10 to 38 in each subgroup from experimental group 3; (**b**) corresponding tumor progression ratio determined from (**a**) for a time period of 7 days [59]

and can be used to measure the degree of stenosis in arteries and monitor the perfusion rates of tumors and organs after transplantation [57].

However, despite all of the promising preclinical data, microbubbles are only clinically utilized for imaging purposes. Currently, physician acceptance and government approval are the two most significant obstacles faced by microbubble technologies, noting that regulatory agency approval must be satisfied if the use of microbubbles is to expand [56].

Effects of Focused Ultrasound (FUS) and Microbubbles on Blood-Brain Barrier (BBB): Due to the heterogeneous permeability of brain tumors, the delivery of therapeutic agents across the blood-brain barrier (BBB) remains a key challenge that significantly hinders the clinical effectiveness of current treatments. The combination of FUS with the intravenous administration of microbubbles has successfully demonstrated the transient opening of the BBB and thereby increasing the chemotherapeutic drug dosage localized to brain tumors in human glioma cell-bearing mice models. Specifically, FUS-BBB opening increased the local accumulation of TMZ (temozolomide – the current standard of care and most important chemotherapy agent administered to control glioma progression) in the brain from 6.98 to 19 ng/mg and reduced the detrimental effects of systemic toxicity due to enhanced targeting. As shown in Fig. 16, researchers improved tumor progression and mean survival rates (from 36 ± 6.9 to 75.1 ± 5.7 days), providing evidence for the potential clinical application of this method to improve current brain tumor treatment [60].

Applications of Low-Intensity Pulsed Ultrasound (LIPUS) in Bone Tissue Engineering: US has been documented to improve stimulation of bone regeneration and bio-apsorption, and it is an easy-to-use noninvasive therapy that can be applied to diminish abnormal healing in fractures. Low-intensity pulsed LIPUS has been shown to promote bone formation and resorption in vitro through influencing all major cell types involved in bone healing, including osteoblasts, osteoclasts, chondrocytes, and mesenchymal stem cells [61]. Additionally, the expression of transforming growth factor beta, a protein involved in bone growth and repair, is directly monitored by US [62]. Lastly, US stimulation positively influences bone tissue through biological mechanisms including cellular adhesion, proliferation, differentiation, and gene expression. Although clinical studies, such as an experiment which demonstrated a 38% decrease in the overall healing time of cortical fractures, have been conducted that validate US stimulation applications in bone tissue engineering, more research is required to develop clinical strategies [63].

2.2.4 Future Directions

Current methods of ultrasound therapy such as HIFU show potential and have demonstrated their usefulness in certain clinical scenarios. However, the field of ultrasound is expanding rapidly, and a broad number of US applications are currently being developed which could drastically increase the practicality of US in clinical practice. These innovative techniques utilize previously known techniques in novel ways, as well as completely new methods for ultrasound. Perhaps the most profound new application for US is its use in facilitating the passage of drugs past the blood-brain barrier (BBB). Clinicians and researchers have long sought a way to safely breach the BBB with therapeutics, which under normal circumstances would be incapable of doing so. A new method of using high focused ultrasound directed at a specific brain region, in conjunction with microbubbles, and a fast MRI for monitoring, has proven to be an effective and relatively safe way to resolve the BBB issue. This method of magnetic resonance-guided focused ultrasound (MRgFUS) has proven to be extremely effective and is currently being used to explore different drugs that can now be delivered to the brain, as had never before been possible. Since the discovery of MBs that disrupt the BBB, numerous therapeutics have been experimented with and loaded into or conjugated with MBs to attempt to treat various brain ailments. Perhaps the most advanced and intriguing drug delivery method comes from a group who made MBs multimodal so that they could be used as ultrasound and MRI contrast agents, while at the same time loading these imaging agents with DOX and specific cellular receptor ligands. This agent then can be used not only for multimodal imaging but also for targeted and controlled release of its DOX payload using focused ultrasound. It is this type of innovation that truly is driving the use of US to the forefront of medical research.

One other US technology that is currently being explored to further its potential is the use of acoustic droplet vaporization (ADV), although ADV is already used clinically for vascular visualization, vascular occlusion therapy, molecular targeting, and drug delivery. However, there is one major therapeutic utility for ADV that has not yet reached the clinic or clinical trials. This promising strategy uses ADV to form gas cavities through which enhanced thermal energy can be deposited [11]. ADV in the use for enhanced ablation therapy in one study actually enhanced the lesions caused by HIFU sevenfold, which makes it an extremely potent augmentation to the current HIFU technique [12].

The future of US-mediated therapies, especially therapies which utilize microbubbles for targeted gene and drug delivery, depends on a greater understanding of molecular targeting and microbubble architecture. Drug loading and kinetic release properties of microbubbles in ultrasound fields highly depend on the microbubble shell composition and gas core. Studies have shown increased in vivo transfection efficiency from intramuscular injection of biotinylated microbubbles [64]. Other studies have also shown longer circulation half-lives of cationic microbubbles coated with small peptides (<1 kDa) [65]. Similar techniques may someday be applied to large DNA vectors.

Understanding the effects of microbubble size on circulation life, cavitation, and vascular extravasation is another field of investigation. Studies have shown that that microbubbles of 4–8 μm diameters are successful at opening the BBB at lower pressures than microbubbles <4 μm [66]. Hence, understanding how to optimize particle size and surface chemistry to achieve specific bio-effects will expand the applicability of microbubbles in various therapies.

Finally, molecular targeting for site-specific US therapies is an area of interest. Molecular targeting can decrease microbubble dose and enhance the effects of sonoporation. Studies have demonstrated enhanced in vitro sonoporation; however, studying the efficacy in vivo may someday expand microbubble applications into clinical settings. Cutting-edge US therapy is a major scientific and engineering marvel. The applications of US through HIFU, microbubbles, molecular targeting, ADV, and other methods are extremely useful in the clinic and laboratory. It is safe to say that US will only be growing as a critical part of healthcare profession, for its wide range of utility in diagnostics, and more recently therapeutics.

3 Conclusion

This introductory chapter is devoted broadly to the topic of theranostics and the role of functional nanometer-sized agents in personalized medicine [1, 50, 59, 67–74]. Over the past 2 decades, the field has gained a tremendous boost with high impending for translation. Advancements in the areas of chemistry, molecular biology, genetics, and engineering created opportunities for interdisciplinary with the objective of driving medical imaging and therapeutic strategies for early, sensitive detection, diagnosis, and treatment of a disease at the molecular and cellular level with uncompromised specificity [5, 7, 9, 75, 76]. A myriad of advancements has been made toward the development of defined nanostructures for performing dual function, i.e., imaging and therapy. However, their clinical translation is still far-reaching. Better understanding of biological and biophysical obstacles encountered by these agents is necessary [11, 12, 14, 22, 72, 77, 78]. For the readers, this introductory chapter will illustrate a presentation of the advancements related to this field and the biological obstacles encountered, which we hope will stimulate more studies to tune these technologies for translational and clinical applications.

References

1. Shin SJ, Beech JR, Kelly KA (2012) Targeted nanoparticles in imaging: paving the way for personalized medicine in the battle against cancer. Integr Biol (Camb) 5(1):29–42
2. Pang T (2012) Theranostics, the 21st century bioeconomy and 'one health'. Expert Rev Mol Diagn 12(8):807–809
3. Lee DY, Li KC (2011) Molecular theranostics: a primer for the imaging professional. AJR Am J Roentgenol 197(2):318–324
4. James ML, Gambhir SS (2012) A molecular imaging primer: modalities, imaging agents, and applications. Physiol Rev 92(2):897–965
5. Pan D, Lanza GM, Wickline SA, Caruthers SD (2009) Nanomedicine: perspective and promises with ligand-directed molecular imaging. Eur J Radiol 70(2):274–285
6. Alberti C (2012) From molecular imaging in preclinical/clinical oncology to theranostic applications in targeted tumor therapy. Eur Rev Med Pharmacol Sci 16(14):1925–1933
7. Wang LS, Chuang MC, Ho JA (2012) Nanotheranostics – a review of recent publications. Int J Nanomedicine 7:4679–4695
8. Lammers T, Aime S, Hennink WE, Storm G, Kiessling F (2011) Theranostic nanomedicine. Acc Chem Res 44(10):1029–1038
9. Cabral H, Nishiyama N, Kataoka K (2011) Supramolecular nanodevices: from design validation to theranostic nanomedicine. Acc Chem Res 44(10):999–1008
10. Kunjachan S, Jayapaul J, Mertens ME, Storm G, Kiessling F, Lammers T (2012) Theranostic systems and strategies for monitoring nanomedicine-mediated drug targeting. Curr Pharm Biotechnol 13(4):609–622
11. Pan D, Caruthers SD, Chen J, Winter PM, SenPan A, Schmieder AH, Wickline SA, Lanza GM (2010) Nanomedicine strategies for molecular targets with MRI and optical imaging. Future Med Chem 2(3):471–490
12. Lanza GM (2012) ICAM-1 and nanomedicine: nature's doorway to the extravascular tissue realm. Arterioscler Thromb Vasc Biol 32(5):1070–1071
13. Pan D, Pramanik M, Senpan A, Ghosh S, Wickline SA, Wang LV, Lanza GM (2010) Near infrared photoacoustic detection of sentinel lymph nodes with gold nanobeacons. Biomaterials 31(14):4088–4093
14. Albanese A, Tang PS, Chan WC (2012) The effect of nanoparticle size, shape, and surface chemistry on biological systems. Annu Rev Biomed Eng 14:1–16
15. Elsabahy M, Wooley KL (2012) Design of polymeric nanoparticles for biomedical delivery applications. Chem Soc Rev 41(7):2545–2561
16. Popovtzer R, Agarwal A, Kotov NA, Popovtzer A, Balter J, Carey TE, Kopelman R (2008) Targeted gold nanoparticles enable molecular CT imaging of cancer. Nano Lett 8(12):4593–4596
17. Schlomka JP, Roessl E, Dorscheid R, Dill S, Martens G, Istel T, Bäumer C, Herrmann C, Steadman R, Zeitler G, Livne A, Proksa R (2008) Experimental feasibility of multi-energy photon-counting K-edge imaging in pre-clinical computed tomography. Phys Med Biol 53 (15):4031–4047
18. Brenner DJ (2010) Should we be concerned about the rapid increase in CT usage? Rev Environ Health 25(1):63–68
19. Hamlin DJ, Burgener FA (1981) Positive and negative contrast agents in CT evaluation of the abdomen and pelvis. J Comput Tomogr 5(2):82–90
20. Lanza GM, Pan D (2014) Molecular imaging with computed tomography. Contrast Media Mol Imaging 9(1):1–2
21. Lee N, Choi SH, Hyeon T (2013) Nano-sized CT contrast agents. Adv Mater 25(19):641–660
22. Pan D (2015) Next generation gene delivery approaches: recent progress and hurdles. Mol Pharm 12(2):299–300

23. Pan D, Schirra CO, Wickline SA, Lanza GM (2014) Multicolor computed tomographic molecular imaging with noncrystalline high-metal-density nanobeacons. Contrast Media Mol Imaging 9(1):13–25
24. Liu Y, Ai K, Lu L (2012) Nanoparticulate X-ray computed tomography contrast agents: from design validation to in vivo applications. Acc Chem Res 45(10):1817–1827
25. Pysz MA, Gambhir SS, Willmann JK (2010) Molecular imaging: current status and emerging strategies. Clin Radiol 65(7):500–516
26. Piechowiak EL et al (2015) Intravenous iodinated contrast agents amplify DNA radiation damage at CT. Radiology 275(3):692–697
27. Kim D et al (2007) Antibiofouling polymer-coated gold nanoparticles as a contrast agent for in vivo computed tomography imaging. J Am Chem Soc 129(24):7661–7665
28. Weber WA et al (2008) Technology insight: advances in molecular imaging and an appraisal of PET/CT scanning. Nat Clin Pract Oncol 5:160–170
29. Reuveni T et al (2011) Targeted gold nanoparticles enable molecular CT imaging of cancer: an in vivo study. Int J Nanomedicine 6:2859–2864
30. Shilo M et al (2012) Nanoparticles as computed tomography contrast agents. Nanomedicine 7 (2):257–259
31. Kao HW et al (2014) Biological characterization of cetuximab-conjugated gold nanoparticles in a tumor animal model. Nanotechnology 25(29):1
32. Mieszawska AJ et al (2013) Multifunctional gold nanoparticles for diagnosis and therapy of disease. Mol Pharm 10(3):831–847
33. Kao HW et al (2013) Evaluation of EGFR-targeted radioimmuno-gold-nanoparticles as a theranostic agent in a tumor animal model. Bioorganic Med Chem Lett 23(11):3180–3185
34. Rabin O et al (2006) An X-ray computed tomography imaging agent based on long-circulating bismuth sulphide nanoparticles. Nat Mater 10(10):118–122
35. Pan D et al (2010) Computed tomography in color: NanoK-enhanced spectral CT molecular imaging. Angew Chem Int Ed Engl 49(50):9635–9639
36. Clark DP, Badea CT (2014) Micro-CT of rodents: state-of-the-art and future perspectives. Phys Med 30(6):619–634
37. Peters S, Stahel RA (2014) Successes and limitations of targeted cancer therapy. In: Progress in tumor research, vol 41. Karger Medical and Scientific Publishers, Basel
38. Tapfer A, Braren R, Bech M, Willner M, Zanette I et al (2013) X-Ray phase-contrast CT of a pancreatic ductal adenocarcinoma mouse model. PLoS One 8(3), e58439
39. Hernot S, Klibanov AL (2008) Microbubbles in ultrasound-triggered drug and gene delivery. Adv Drug Deliv Rev 60(10):1153–1166
40. Cosgrove D (2006) Ultrasound contrast agents: an overview. Eur J Radiol 60(3):324–330
41. Sirsi SR, Borden MA (2009) Microbubble compositions, properties and biomedical applications. Bubble Sci Eng Technol 1(1–2):3–17
42. Lefor AT et al (1995) The effects of hyperthermia on vascular permeability in experimental liver metastasis. J Surg Oncol 28(4):297–300
43. May J, Li SD (2013) Hyperthermia-induced drug targeting. Expert Opin Drug Deliv 10 (4):511–527
44. Frenkel V et al (2006) Delivery of liposomal doxorubicin (doxil) in a breast cancer tumor model: investigation of potential enhancement by pulsed-high intensity focused ultrasound exposure. Acad Radiol 13(4):469–479
45. Wang T et al (2014) Ultrasound and microbubble guided drug delivery: mechanistic understanding and clinical implications. Curr Pharm Biotechnol 14(8):743–752
46. Hynynen K, McDannold N, Vykhodtseva N, Jolesz FA (2001) Noninvasive MR imaging-guided focal opening of the blood-brain barrier in rabbits. Radiology 220(3):640–646
47. Society of Diagnostic Medical Sonography (2015) Sound medicine: understanding ultrasound and its benefits. Society of Diagnostic Medical Sonography
48. Miller D, Smith N, Bailey M, Czarnota G, Hynynen K, Makin I (2012) Overview of therapeutic ultrasound applications and safety considerations. J Ultrasound Med 31(4):623–634

49. Miller DL (1987) A review of the ultrasonic bioeffects of microsonation, gas-body activation, and related cavitation-like phenomena. Ultrasound Med Biol 13(8):443–470
50. Newman CM, Bettinger T (2007) Gene therapy progress and prospects: ultrasound for gene transfer. Gene Ther 14:465–475
51. Robertson VJ, Baker KG (2001) A review of therapeutic ultrasound: effectiveness studies. Phys Ther 81(7):1339–1350
52. Wood AKW, Sehgal CM (2015) A review of Low-intensity ultrasound for cancer therapy. Ultrasound Med Biol 41(4):905–928
53. Neisius A et al (2015) Shock wave lithotripsy: the new phoenix? World J Urol 33(2):213–221
54. Grippaudo FR et al (2004) Ultrasound-assisted liposuction for the removal of siliconomas. Scand J Plast Reconstr Surg Hand Surg 38(1):21–26
55. Shridharani SM, Broyles JM, Matarasso A (2014) Liposuction devices: technology update. Med Devices Evid Res 7:241–251
56. Martin KH, Dayton PA (2013) Current status and prospects for microbubbles in ultrasound theranostics. Wiley Interdiscip Rev Nanomed Nanobiotechnol 5(4):329–345. doi:10.1002/wnan.1219 PM
57. Kang ST, Yeh CK (2012) Ultrasound microbubble contrast agents for diagnostic and therapeutic applications: current status and future design. Chang Gung Med J 35(2):125–139
58. Sirsi SR, Borden MA (2012) Advances in ultrasound mediated gene therapy using microbubble contrast agents. Theranostics 2(12):1208–1222
59. Liu HL, Fan CH, Ting CY, Yeh CK (2014) Combining microbubbles and ultrasound for drug delivery to brain tumors: current progress and overview. Theranostics 4(4):432–444
60. Liu HL et al (2014) Pharmacodynamic and therapeutic investigation of focused ultrasound-induced blood-brain barrier opening for enhanced temozolomide delivery in glioma treatment. PLoS One 9(12), e114311
61. Claes L, Willie B (2007) The enhancement of bone regeneration by ultrasound. Prog Biophys Mol Biol 93:384–398
62. Zhou S, Schmelz A, Seufferlein T, Li Y, Zhao J, Bachem MG (2004) Molecular mechanisms of low intensity pulsed ultrasound in human skin fibroblasts. J Biol Chem 279:54463–54469
63. Yang MH et al (2010) Application of ultrasound stimulation in bone tissue engineering. Int J Stem Cells 3(2):74–79
64. Nomikou N, Tiwari P, Trehan T, Gulati K, McHale AP (2011) Studies on neutral, cationic and biotinylated cationic microbubbles in enhancing ultrasound-mediated gene delivery in vitro and in vivo. Acta Biomater 8(3):1273–1280
65. Sirsi SR, Hernandez SL, Zielinski L, Blomback H, Koubaa A, Synder M et al (2012) Polyplex-microbubble hybrids for ultrasound-guided plasmid DNA delivery to solid tumors. J Control Release 157:224–234
66. Tung YS, Vlachos F, Feshitan JA, Borden MA, Konofagou EE (2011) The mechanism of interaction between focused ultrasound and microbubbles in blood-brain barrier opening in mice. J Acoust Soc Am 130:3059–3067
67. Chadderdon SM, Kaul S (2010) Molecular imaging with contrast enhanced ultrasound. J Nucl Cardiol 17(4):667–677
68. Lin CY, Pitt WG (2013) Acoustic droplet vaporization in biology and medicine. Biomed Res Int 2013:404361
69. Zhang M, Fabiilli ML, Haworth KJ et al (2011) Acoustic droplet vaporization for enhancement of thermal ablation by high intensity focused ultrasound. Acad Radiol 18(9):1123–1132
70. Kiessling F et al (2014) Recent advances in molecular, multimodal and theranostic ultrasound imaging. Adv Drug Deliv Rev 72:15–27
71. Dobrucki LW, Pan D, Smith AM (2015) Multiscale imaging of nanoparticle drug delivery. Curr Drug Targets 6(6):560–570
72. Pan D, Kim B, Hu G, Gupta DS, Senpan A, Yang X, Schmieder A, Swain C, Wickline SA, Tomasson MH, Lanza GM (2015) A strategy for combating melanoma with oncogenic c-Myc inhibitors and targeted nanotherapy. Nanomedicine (Lond) 10(2):241–251

73. Zhang R, Pan D, Cai X, Yang X, Senpan A, Allen JS, Lanza GM, Wang LV (2015) Avβ_3-targeted copper nanoparticles incorporating an Sn 2 lipase-labile fumagillin prodrug for photoacoustic neovascular imaging and treatment. Theranostics 5(2):124–133
74. Misra SK, Ye M, Kim S, Pan D (2014) Highly efficient anti-cancer therapy using scorpion 'NanoVenin'. Chem Commun (Camb) 50(87):13220–13223
75. Misra SK, Ye M, Kim S, Pan D (2015) Defined nanoscale chemistry influences delivery of peptido-toxins for cancer therapy. PLoS One 10(6), e0125908
76. Mukherjee P, Misra SK, Gryka MC, Chang HH, Tiwari S, Wilson WL, Scott JW, Bhargava R, Pan D (2015) Tunable luminescent carbon nanospheres with well-defined nanoscale chemistry for synchronized imaging and therapy. Small. doi:10.1002/smll.201500728
77. Soodgupta D, Pan D, Cui G, Senpan A, Yang X, Lu L, Weilbaecher KN, Prochownik EV, Lanza GM, Tomasson MH (2015) Small molecule MYC inhibitor conjugated to integrin-targeted nanoparticles extends survival in a mouse model of disseminated multiple myeloma. Mol Cancer Ther 14(6):1286–1294
78. Misra SK, Jensen TW, Pan D (2015) Enriched inhibition of cancer and stem-like cancer cells via STAT-3 modulating niclocelles. Nanoscale 7(16):7127–7132

Top Med Chem (2016) 20: 105–120
DOI: 10.1007/7355_2015_97

Published online: 15 October 2015

Nano-Enabled Delivery of Intracellular Therapeutics

Fatemeh Ostadhossein, Enrique Alejandro Daza, Daniel Frankowski, Drew Goatz, Molly Imgruet, Joseph Kus, Ryan Lake, Mallika Modak, Nick Olsen, Aaron Schwartz-Duval, Alyssa Zimmer, Nicholas Kolmodin, and Dipanjan Pan

Abstract Many diseases that plague the modern medical world have their origins at the cellular or molecular level and, as such, require greater specificity to be effectively combated and cured. A number of recent advances in understanding the biology and biochemistry have enabled researchers to develop the specialized tools and techniques needed to detect and provide therapy for these debilitating conditions. Many of these treatments take advantage of the way that cells behave and interact with their environment or various properties of the cell's structure and form. Researchers are able to surpass a number of cellular hurdles, such as the cell membrane, endosomal escape, and intracellular targeting to begin the arduous task of understanding, diagnosing, and treating diseases like cancer.

Keywords Cell-penetrating peptides, Intracellular delivery, Nanoparticles, Therapeutics

---

F. Ostadhossein, E.A. Daza, D. Frankowski, D. Goatz, M. Imgruet, J. Kus, M. Modak, N. Olsen, A. Schwartz-Duval, A. Zimmer, and N. Kolmodin
Department of Bioengineering, College of Engineering, University of Illinois at Urbana Champaign, Urbana, IL, USA

R. Lake
Department of Chemistry, College of Engineering, University of Illinois at Urbana Champaign, Urbana, IL, USA

D. Pan (✉)
Department of Bioengineering, University of Illinois at Urbana Champaign, Urbana, IL, USA

Beckman Institute for Science and Technology, Urbana, IL, USA

Department of Materials Science and Engineering, University of Illinois at Urbana Champaign, Urbana, IL, USA

Carle Foundation Hospital, Urbana, IL, USA
e-mail: dipanjan@illinois.edu

Contents

Abbreviations

ATP	Adenosine triphosphate
CIE	Clathrin-independent endocytosis
CLIC	Clathrin- and dynamin-independent carriers
CME	Clathrin-mediated endocytosis
CPP	Cell penetrating peptides
CPT	Camptothecin
DNP	Dinitrophenol
ECM	Extracellular matrix
EGFR	Epidermal growth factor receptors
ER	Endoplasmic reticulum
ETC	Electron transport chain
FR	Fc receptor
GUV	Giant unilamellar vesicle
HP	Hematoporphyrin
NLS	Nuclear localization sequence
NP	Nanoparticle
NPC	Nuclear pore complex
PC	Phophatidylcholine
PCI	Photo-chemical internalization
PE	Phosphatidylethanolamine
PEG	Poly-ethylene glycol
PS	Phosphatidylserine
PSMA	Prostate-specific membrane antigens
RES	Reticulo-endothelial system
ROS	Reactive oxygen species
TCHD	*Trans*-cyclohexane-1,2-diol
TIM	Transporter inner membrane
TOM	Transporter outer membrane
TPP	Triphenylphosphonium
TR	Transferrin receptors

1 Introduction

For predominantly fatal and life-shortening diseases in modern society such as Alzheimer's, diabetes, and cancer, the paradigm shift in treatment has been toward the cellular and molecular levels and away from systems and tissue levels. This shift is largely due to the many discoveries of the molecular origins for these specific disease pathways [1–3] which include hereditary risk factors, mutations, and result changes in molecular pathways, which cannot be identified or treated until their pathological effects are on the systemic or tissue levels. Additionally, since current medical capabilities are unable to target, locate, or treat these cellular events effectively, these diseases remain largely undetected until they have progressed to tissue level. This shortcoming can potentially lead to disease spread among other tissue systems and thus shortens the patient's life.

Advances in nanotechnology have allowed for potentially earlier identification and treatment of pathologies that are of cellular and molecular origin ex vivo. Nanoparticles come in various forms such as soft particles (e.g., liposomes [4, 5] dendrimers [6], polymers [7]), hard particles (e.g., quantum dots [8], gold [9, 10], magnetite [11]), or naturally occurring species (e.g. proteins [12], micelles [13, 14], viral envelopes [15]). Compared to molecular or capsule drug emission therapeutic delivery, nanoparticles can deliver higher local concentrations of cytotoxic drug with minimal systemic concentrations [7, 16, 17]. Current nanoparticle-based treatments are capable of combining modality-specific imaging contrast with high drug payload and large surface area targeting ligands for an advanced multipurpose therapy agent. The combinatory relationship between treatment and localization of disease models is exclusively exploited in the nanomedicine field with the new approach for combating disease models known as "theranostics" (a portmanteau of therapy and diagnostics). Theranostic nanoparticles allow for a more appropriate application of personalized medicines as the imaging contrast provided allows the researcher or clinician to track the efficacy of the therapy throughout the application. This personalization with concurrent monitoring of medical treatment becomes especially critical when considering diseases that are largely heterogeneous in nature such as cancer, whose current treatments are associated with emaciation and suffering, almost as highly as the disease.

Although nanoparticles have huge potential in molecular medicine, drug delivery optimization and cellular targeting are bottlenecks in their efficient exploitation. These barriers include human efficiency, such as cost; external barriers, such as skin or mucosa; en route efficiency, such as blood; and cellular barriers that must be overcome in order for a treatment to be successful. Nanomedicine offers solutions to the problems presented by cellular barriers, which offer some of the most varied and difficult challenges in drug delivery, as well as many of the most promising methods for future drug delivery approaches.

2 Crossing the Cell Membrane and Internalization

A primary barrier preventing successful cellular delivery is the cellular membrane. This membrane is composed of a phospholipid bilayer with embedded proteins selectively permeable for ions and organic molecules and is crucial for cell communication and adhesion. Successful translocation across this membrane is critical for further intracellular drug targeting. Endocytosis, the formation of new cytosolic membrane-bound vesicles from the cell plasma membrane, is the primary method of internalization of extracellular components (Fig. 1). The two principal endocytotic pathways utilized by cells are phagocytosis and pinocytosis. Phagocytosis is used by a multitude of cell types to engulf foreign particles as part of the immune response. Interaction of cell-surface receptors with factors that recognize the foreign body or with the foreign body itself triggers phagocytosis. Receptors that have been identified as facilitating phagocytosis include the Fc receptor (FR) family and complement receptors [19].

In the case of nanoparticles, attractive forces such as van der Waals, electrostatic, ionic, and hydrophobic/hydrophilic between nanoparticles and cells facilitate internalization via phagocytosis [20, 21]. These forces are affected by the contact angle between the nanoparticle and host cell membrane [21]. Differences in nanoparticles' geometry have been shown to affect the success of internalization

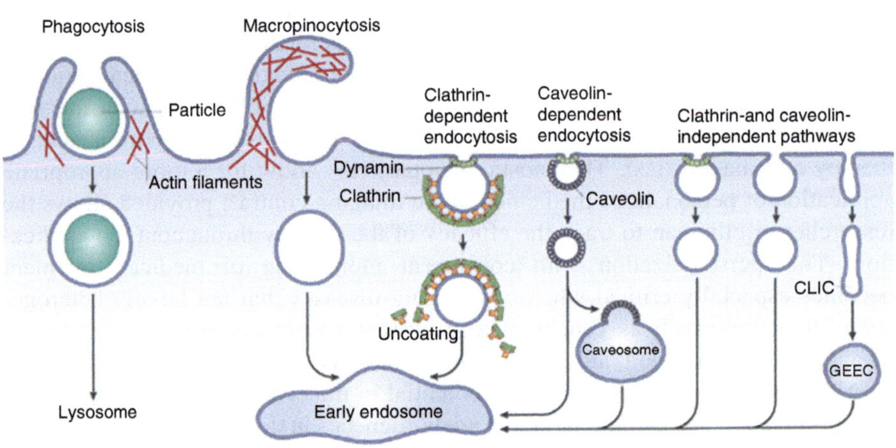

Fig. 1 Pathways of entry into cells. Large particles can be taken up by phagocytosis, whereas fluid uptake occurs by macropinocytosis. Numerous cargoes can be endocytosed by mechanisms that are independent of the coat protein clathrin and the fission GTPase dynamin. Most internalized cargos are delivered to the early endosome via vesicular (clathrin- or caveolin-coated vesicles) or tubular intermediates known as clathrin- and dynamin-independent carriers (CLICs) that are derived from the plasma membrane. Some pathways may first traffic to intermediate compartments, such as the caveosome or glycosylphosphatidylinositol-anchored protein-enriched early endosomal compartments (GEEC), en route to the early endosome. Reproduced with permission from [18]

via phagocytosis, due to the varying contact angles at the cell membrane surface caused by different particle shapes [22]. In a comparison of nanoparticles of various shapes and aspect ratios, it was found that particles that were elongated with higher aspect ratios were less likely to be internalized via phagocytosis [23]. Concurrently, a similar study found that particles with higher aspect ratios were more prone to endosomal and lysosomal localization [24]. Nanoparticle size and shape tunability is thus an important tool in developing targeted nanomedicines, but must be carefully controlled in order to achieve the desired outcome, whether it is phagocytosis or specific intracellular targeting [25]. Modulation of particle properties also has been shown to affect internalization via pinocytosis. Pinocytosis is clathrin mediated (CME), clathrin independent (CIE), or caveolae mediated [26]. Nanoparticles can be made more susceptible to these internalization pathways by modulating size, shape, and surface charge. Positively charged nanoparticles have been shown to be preferentially taken up through CME, while particles with negative surface charges are associated with internalization via caveolae [27, 28].

Nanomedicine presents an attractive option because it has no cargo size limitations and can specifically be targeted to certain cellular receptors [29]. Carbon nanoparticles have been identified as possible carriers of DNA molecules and have shown a high transfection efficacy in breast cancer cells, as shown in Fig. 2 [30]. Nanoparticles have the potential to be effective carriers for a large variety of different materials which help to increase cargo uptake by the cells. Additionally, when compared to delivering small molecule drugs alone, nanoparticles can increase delivery efficiency, leading to lower effective dosages and fewer side effects [31].

More recently, dendrimers have been shown to function as effective intracellular carriers for therapeutic and imaging agents. The new generation of dendrimer-based delivery systems has shown to be capable of bypassing efflux transporters to enable the efficient transport of drugs across cellular barriers.

Receptor-mediated endocytosis is an internalization method that is used to deliver nanoparticles to disease sites by exploiting the overexpression of cell-surface receptors on target disease cells. This method of active targeting has been utilized for the delivery of both small molecule drugs and nucleic acids and is achieved via the functionalization of nanoparticle surfaces with targeting ligands including small molecules, peptides, antibodies, and aptamers. In the context of tumor targeting, folate receptors (FR), epidermal growth factor receptors (EGFR),

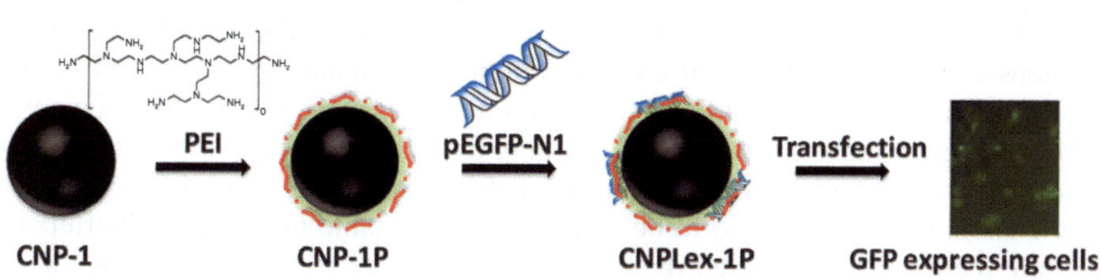

Fig. 2 Carbon nanoparticles used for gene delivery. Image of cells transfected with the pEGFP-N1 reporter gene plasmid in breast cancer cell line MDA-MB231. Reproduced with permission from [30]

transferrin receptors (TR), prostate-specific membrane antigens (PSMA), and integrins have been implicated in different types of cancer and thus used as targeting ligands in order to specifically deliver therapeutic nanoparticles to tumor sites with minimal off-target toxicity [32].

2.1 Endosomal Escape and Cytosolic Delivery

After the payload has been successfully internalized, it must still pass the subcellular obstacles such as the early endosome, late endosome, and lysosome. This critical moment in the subcellular delivery of nanoparticles can either result in lysosomal degradation, exocytotic release, or trafficking of particles to the desired organelle [33]. Specifically, it is of paramount importance that nanoparticles escape the endosome because vesicular sequestration impedes delivery of the cargo and leads to degradation of the nanoparticle [23]. Vesicular entrapment is widely regarded as an undesirable phenomenon, unless targeting lysosomal storage disorder. Several strategies have been developed to circumvent vesicular entrapment such as fusogenic peptides [34, 35], pH-sensitive polymers, pH-sensitive core shell nanoparticles [36], and pH-sensitive liposomes. Cationic liposomes, polypeptides, amine-containing polymers, and cationic lipids have been shown to be efficient in non-viral gene therapy. These materials interact electrostatically with membrane glycoproteins, proteoglycans, or other anionic membrane components efficaciously as non-viral vectors [34, 37, 38].

2.2 Cationic Escape

There are two methods that enable cationic materials to undergo endosomal escape. One strategy involves the material's interaction with endosomal membrane and subsequent pore formation facilitating the transport to the cytosol. A second cationic endosomal escape strategy utilizes the "proton sponge effect" during which the endosomal membrane ruptures and the cargo is released directly into the cytosol [10]. Through endosomal maturation, the pH significantly decreases from 6 to 4 and an excess of protons can be sequestered by the contribution of the protons from amine groups in cationic polymers, maintaining the action of the proton pumps. A parallel influx of Cl^- and water takes place so as to keep a neutral pH of the environment, resulting in swelling and subsequent rupture of the endosome [33]. Although a viable platform for direct release of cargo into cytosol is protonated, they are claimed to be cytotoxic and unstable in biological buffers or culture media and are cleared rapidly upon exposure to the extracellular environment by the reticuloendothelial system (RES) [33]. These concerns have partially been resolved via surface passivation by materials such as polyethylene glycol (PEG), dextran, Pluronics, and human serum albumin [39]. In addition, there exists a complementary method called photochemical internalization (PCI) during which a photosensitizing molecule conjugated with drug is photochemically illuminated, subsequently triggering the formation of reactive oxygen species (ROS) and ultimately causing endosomal rupture [40]. However, this method has some limitations such as potential damage to the drug due to singlet oxygen exposure. Coupling of PCI with pH-responsive systems in which photosensitizing agents can become active only in low pH has been utilized to enhance the overall efficacy [41]. For instance, very recently, Pasparakis et al. [42] developed a novel self-assembling polymer of the polyacetal family, which is degradable by light and pH. They used this polymer to indicate the potential of photochemical internalization in a multimodal therapy approach combining chemo- and photothermal therapy. The phototoxic drug hematoporphyrin (HP) and the chemotherapeutic anticancer agent camptothecin (CPT) were incorporated within the polymeric nanoparticles which can subsequently be activated using visible wavelength leading to cancer cell death due to light and pH-mediated intracellular delivery of drug payload. The polymer was synthesized via acid-catalyzed polycondensation reaction of 2-nitroresorcinol and cyclohexyl divinyl ether which was further capped by poly(ethylene glycol). The spherical particles had hydrodynamic size of 190 nm and were found to be stable in slightly alkaline solutions for weeks. The CPT release profile of the polymer under both acidic (pH 5.2) and light irradiation condition was significantly enhanced (>90%) compared to non-irradiative condition (~52%), indicating the role of HP in generating the reactive oxygen species. Furthermore, preliminary cytotoxicity studies on HeLa cells revealed that the drugs acted more effectively in the samples under both irradiative and low pH conditions compared to the non-irradiative case (death rate of 52% vs 27%). In addition, fluorescence microscopy investigation of the developed nanoparticles confirmed the uptake of nanoparticles by strong absorption at characteristic absorption of HP at 400 nm. The authors contended that the mechanism for cellular uptake consists of NP endocytosis pathway and translocation to the late endosome where the cargo gets hydrolyzed and the effect is further boosted due to laser photolysis leading to endosomal degradation and release of CPT. Overall, this study utilizes clever chemistry alongside with nanoparticle approach to cross the endosome compartment through pH and ROS generation at visible range as two dominant factors. Figure 3 summarizes the role of CPI as a viable method to cross the endosome barrier.

Fusogenic Peptide Escape: Some synthetic peptides containing fusogenic peptides (such as GALA99 or KALA sequences) are also capable of enhancing endosomal escape. At physiological pH, these peptides coil as they are rich in anionic carboxyls, while they form α-helix secondary structure upon protonation at a lower pH, such as inside the endosome. This α-helix secondary structure can interact with and destabilize the lipid bilayer, leading to endosomal escape [43–45].

Cell-Penetrating Peptides: Another strategy for endosomal escape is the use of cell-penetrating peptides (CPPs) which facilitate the translocation of cargo along the membrane and make the direct release of the drug into the cytosol feasible [46, 47]. Despite being studied extensively, their mechanism of traversing the membrane remains highly elusive. Studies suggest that the interaction of CPPs'

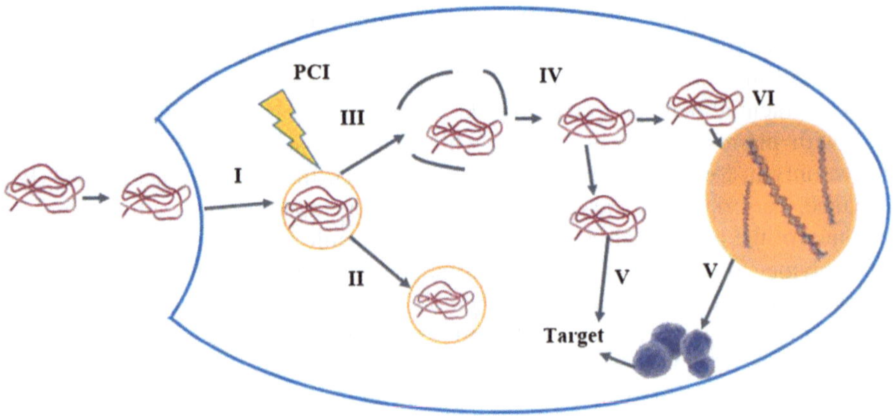

Fig. 3 Photochemical internalization pathway. (I) Endocytosis, (II) light exposure, and singlet oxygen generation (III) rupture of vesicular membrane due to oxidative damage (IV) release of the payload into the cytosol which can be either targeted to (V) cytoplasm or (VI) nucleus leading to (V) transgene activation. Alternative route is (II) hydrolytic degradation by endosome and lysosome

cationic lipid region with phospholipid membrane and conformational changes can facilitate the lipid head insertion, while other studies refer to endocytosis as the dominant mode of internalization [48–50]. The current scientific understanding is that CPPs induce various types of endocytosis using some of their physicochemical properties, such as molecular length, charge delocalization, and hydrophobicity. CPPs have gained much attention recently and are currently being investigated in preclinical studies, where they have shown to be successful for helping to address a wide variety of conditions [51]. It should be acknowledged that the low specificity of CPP is the main limiting factor in their application. This low specificity has been remedied through conjugation with other more specific ligands. Furthermore, to boost their efficacy and ameliorate the cytotoxic effects, modification with fatty acids (such as cholesterol, cysteamine, CPP-like ligands, and various guanidine-rich transporters) has been investigated [45]. For example, the TAT peptides derived from HIV1 have the ability to penetrate the cell membrane and deliver cargoes into the cytoplasm without endosomal or lysosomal degradation. TAT proteins have been effective at delivering a variety of molecular cargoes, including proteins with a mass greater than 100 kDa, 40 nm nanoparticles, and 200 nm liposomes [52, 53]. TAT proteins also have the potential to generate pores in model membranes. Giant unilamellar vesicles (GUVs) were constructed as model membranes that were made of only phophatidylcholine (PC), PC and phosphatidylserine (anionic) (PS), or PC and phosphatidylethanolamine (cationic) (PE). Each membrane also contained cholesterol to better mimic physiological membranes. TAT was effectively able to translocate across both the PC/PS and PC/PE membranes, but not the PC alone. Each membrane had a different interaction with TAT based on the charges present in the membrane. In PC/PS GUVs, these interactions would cause the GUVs to deform after 20–30 min, and they would eventually rupture, releasing their contents. In PC/PE GUVs, these interactions were only seen at 20% and 30% PE after 30 min, but not at 10%. The GUVs never burst when the membrane composition was PC/PE. This study showed that TAT peptides accumulated on anionic membranes and were very rapidly internalized by the GUVs. It was also observed that these peptides were able to translocate across membranes containing lipids that induce negative curvature to the membrane such as PE [54].

2.3 pH-Sensitive Liposomes

pH-sensitive liposomes are designed to be endocytosed, but facilitate lysosomal escape of their drug cargo upon acidification during endosomal maturation. The exact mechanism for drug lysosomal escape to the cytoplasm via pH-sensitive liposomes is unknown, but theories include liposome-facilitated destabilization of the lysosomal membrane, passive diffusion across the lysosomal membrane, and pH-triggered fusion of liposomal and lysosomal membranes [55]. As a recent example, Turk et al. [56] developed folate-targeted liposomes incorporating pH-sensitive peptides. The peptide was designed with specific arrangement of hydrophobic and hydrophilic amino acid residues to disrupt the liposomal membrane at lower peptide concentrations than previously used peptides. At neutral pH, the peptides are in a mostly random coil conformation; upon acidification to pH values of around 5, the peptides adopt an amphipathic alpha helical structure. This structural change allows the peptides to insert themselves into membranes in a cooperative, self-aggregating manner, inducing permeabilization of the liposomal and subsequently lysosomal membranes. When loaded within these pH-sensitive liposomes, cytosine arabinoside showed a 30-fold increase in potency compared to the free drug [56].

2.4 Intracellular Targeting

In delivering drug within the cell, the cytoplasm acts as an additional barrier which the drug must overcome. This barrier presents itself in two ways. The first is the degradation that may occur as a particle passes through the cytoplasm, and the second is the route that has to be taken to get from one place to another in the cytoplasm.

Cells use the ubiquitin proteasome pathway to degrade proteins in the cytoplasm. This can pose an issue in drug delivery if proteins in the drug delivery particle are marked for degradation. This can degrade all or part of the drug, rendering it ineffective, or it can destroy the nanoparticle, leading to the premature release of the drug before it has reached its final target.

The transport of a drug can also be inhibited by the drug carrier particle's size. Moving through the cytoplasm is only possible passively with smaller particles. This is due to the high density of organelles and macromolecular crowding in the cytoplasm. Larger particles must interact with molecules in order to form a cytoplasmic sieve. This allows the larger molecules to pass through the cytoplasm [57].

Recent research has shown there are ways to avoid hindrance in the transport of drug through the cytoplasm. PEGylation of nanoparticles has been found to decrease the number of particles that are hindered in their transport through the cytoplasm. It has been shown that PEGylation doubles the diffusion rate across the cytoplasm and decreases the amount of hindered particles from 79.2% to 48.8%. It is believed that PEGylation reduces nonspecific adhesion to the cytoskeleton, allowing the nanoparticle to move freely within the cytosol [58].

As mentioned previously, intracellular targeting poses multiple challenges, which can open access to the vast number of highly significant targeting moieties once overcome. Nanomedicine targeting inside human cells has focused on inhibiting or causing a change to natural biochemical reactions contained in organelles or directly within the cytoplasm. The nucleus, mitochondria, lysosome, endoplasmic reticulum, and the Golgi apparatus are popular organelles to study because of the high traffic of cellularly dependent reactions.

The mitochondria serve as the cell's power plant, providing the necessary Adenosine triphosphate (ATP) for many enzymatic reactions and active transport. This double-membrane enveloped organelle is believed to have originated as an extracellular organism which forms a symbiotic relationship with prokaryotes and eukaryotes, thus explaining the existence of its own internal genome.

In addition to ATP synthesis, the mitochondria also play a role in calcium homeostasis regulation and initiation of programmed cell death [59]. Intramitochondrial issues are considered markers for cancer, Parkinson's, Alzheimer's, and amyotrophic lateral sclerosis, thus highlighting the importance of accessing mitochondrial processes for therapeutic nanomedicine [60–63].

The transport proteins transporter inner membrane (TIM) and transporter outer membrane (TOM) provide access to the inner and outer mitochondrial membranes, respectively, and have become attractive ports for drug delivery. Size limitations of these beta-barrel porin-like transport proteins have been reported to restrict passage to molecules smaller than 6 kDa. Once inside the intermembrane space, multiple targeting moieties are open for interaction, for instance, the ATP synthesis factory electron transport chain (ETC). This highly negative system of proteins built into the inner membrane attracts positively charged molecules such as triphenylphosphonium (TPP), dequalinium, or the fluorescent dye rhodamine [52, 64]. Additionally, the protein cytochrome C becomes accessible. As cytochrome C is a crucial component in delivering electrons to the final hydrogen pump, it is directly involved in the apoptosis pathway.

Current commercially available mitochondrial targeting drugs include lonidamine, alpha tocopheryl succinate for cancer, curcumin for Alzheimer's, and Dinitrophenol (DNP) for obesity [4, 7]. Potential future treatments can involve the mitochondrial delivery of antioxidants, proapoptotic factors, drugs, proteins, and nucleic acids [64].

Nuclear Delivery: The nucleus holds the cell's genetic information necessary for protein building which in turn determines the cell function and fate. Targeting this organelle with gene delivery, drugs, or various activators and inhibitors can lead to a multitude of induced therapeutic processes which can be utilized to combat gene-related illnesses. The nucleus is also considered to be one of the most challenging yet significant subcellular organelle targets in nanomedicine. Once a drug or other nanomedicine substance is inside the cell, the next barrier to overcome is the double-membrane nuclear envelope which separates DNA from the cytosol. A well-known strategy for targeting a cell's DNA involves precise timing of cell stage development and delivery. Specifically, the mitotic phase of the cell cycle is where the nuclear envelope breaks down and leaves DNA accessible to cytosolic payloads [65]. Another common approach to cross the nuclear envelope is via the nuclear pore complex (NPC), a receptor-mediated transport protein for RNA and ribosomal proteins as well as a passive diffusion port for the small molecules. The passive diffusion properties have been investigated, and it has been reported that using the amphipathic alcohol trans-cyclohexane-1,2-diol (TCHD) results in pore dilation, effectively increasing the nucleus's passive diffusion ability [66–69].

The specific ligand studied for NPC active transport is a chain of consecutive lysines (or PKKKRKV), also known as the nuclear localization sequence (NLS) which has been taken advantage of and labeled across plasmid DNA and nanoparticles [70]. Karyopherin-beta-mediated transport is an additional method that works as an NLS for different proteins [71]. However, the limitations for transport across the NPC have been reported as 60 kDa (10 nm) [68, 72]. A final approach for crossing the nuclear envelope is through passive diffusion across the lipid membrane, which is governed by the same laws as the main cellular membrane, diffusible only to small molecules and ions [73, 74].

Golgi Apparatus: The Golgi body and the endoplasmic reticulum (ER) are also of great interest for researchers pursuing subcellular-targeted nanoparticles. The Golgi body is associated with Alzheimer's, Parkinson's, and several other lethal congenital diseases. Malfunctions in Golgi body have been linked to prostate cancer. ER mutations have also played a role in diabetes insipidus, chronic pancreatitis, and cancer. Work has begun to target the mTOR pathway, which plays a crucial role in cancer cell growth and which exists mainly in the ER and Golgi body [75]. Viruses have been used to target the nucleus, as well as the ER and the Golgi apparatus. Specifically, the Simian vacuolating virus 40 is particularly adept at targeting these organelles. However, as with all viral-mediated delivery, there is a high risk of toxicity and immune reaction. Nanoparticles outfitted with some of the same sequences and peptides that allow for viral targeting could be very useful in avoiding this immune response but retaining organelle specificity [25].

3 Conclusion

In this review, we have highlighted hurdles in crossing the cellular membrane, endosomal escape, and intracellular targeting. Although nanomedicine and extra-/ intracellular targeting have been studied for over a decade, these hurdles have historically been the limiting factor on nanomedicine achieving clinical implementation. One of the primary hurdles to overcome is improving and optimizing internalization and endosomal escape which is a crucial step before any payload delivery or organelle targeting takes place. Once such a structure is designed, the modalities involved in nanomedicine delivery should be perfected. This includes studying the travel mechanics of the payload within microfluidic-like environments and product interactions with endothelial lining on a three-dimensional plane. Extracellular matrix (ECM) gels and microfluidic platforms are excellent tools for this type of investigation. Once localization within the body is well established, the next significant challenge is the simultaneous optimization of both extracellular and intracellular targeting techniques. Biomimicry of viruses is a great modality to improve the design of synthetic subcellular targeting systems, essentially using a virus as a guide in the design of a nanoparticle. Similarly, surface treatments such as PEGylation may address some of these concerns mentioned, which may otherwise inhibit favorable cellular interactions. Depending on molecular weight, polarity, and surface charge of the nanoparticles, some membrane penetrating routes may be preferred over others.

Once inside the cell, the next challenge is decreasing cytotoxicity while improving therapeutic efficacy, or overcoming additional membranes for organelle-specific targeting. Image capable ligands can also be incorporated in nanomedicine to allow for the visualization of drug transport and action. Future research on the horizon includes the physicochemical characterizations and bioproducts of nanoparticles and clinical determinants in the human body. Despite major advances, there is still significant work ahead to be done, but the progress does not seem to be slowing down and instead is increasing its speed of innovation within the realm of therapeutic nanomedicine.

References

1. Gong Y et al (2003) Alzheimer's disease-affected brain: presence of oligomeric Aβ ligands (ADDLs) suggests a molecular basis for reversible memory loss. Proc Natl Acad Sci 100 (18):10417–10422
2. Nisticò L et al (1996) The CTLA-4 gene region of chromosome 2q33 is linked to, and associated with, type 1 diabetes. Hum Mol Genet 5(7):1075–1080
3. Hemminki A, Hemminki K (2005) The genetic basis of cancer. In: Cancer gene therapy. Springer, Totowa, pp 9–18
4. Mehnert W, Mäder K (2001) Solid lipid nanoparticles: production, characterization and applications. Adv Drug Deliv Rev 47(2):165–196

5. Mukundan S Jr et al (2006) A liposomal nanoscale contrast agent for preclinical CT in mice. Am J Roentgenol 186(2):300–307
6. Pillai O, Panchagnula R (2001) Polymers in drug delivery. Curr Opin Chem Biol 5(4):447–451
7. Jagur-Grodzinski J (2009) Polymers for targeted and/or sustained drug delivery. Polym Adv Technol 20(7):595–606
8. Delehanty JB, Mattoussi H, Medintz IL (2009) Delivering quantum dots into cells: strategies, progress and remaining issues. Anal Bioanal Chem 393(4):1091–1105
9. Pan D et al (2010) Near infrared photoacoustic detection of sentinel lymph nodes with gold nanobeacons. Biomaterials 31(14):4088–4093
10. Ding Y et al (2014) Gold nanoparticles for nucleic acid delivery. Mol Ther 22(6):1075–1083
11. Yue-Jian C et al (2010) Synthesis, self-assembly, and characterization of PEG-coated iron oxide nanoparticles as potential MRI contrast agent. Drug Dev Ind pharm 36(10):1235–1244
12. Ziv K et al (2010) Ferritin as a reporter gene for MRI: chronic liver over expression of h-ferritin during dietary iron supplementation and aging. NMR Biomed 23(5):523–531
13. Pan D et al (2012) Rapid synthesis of near infrared polymeric micelles for real-time sentinel lymph node imaging. Adv Healthc Mater 1(5):582–589
14. Talelli M, Hennink WE (2011) Thermosensitive polymeric micelles for targeted drug delivery. Nanomedicine 6(7):1245–1255
15. Flexman JA et al (2008) Magnetically targeted viral envelopes: a PET investigation of initial biodistribution. IEEE Trans Nanobioscience 7(3):223–232
16. Wang AZ, Langer R, Farokhzad OC (2012) Nanoparticle delivery of cancer drugs. Annu Rev Med 63:185–198
17. Egusquiaguirre SP et al (2012) Nanoparticle delivery systems for cancer therapy: advances in clinical and preclinical research. Clin Transl Oncol 14(2):83–93
18. Mayor S, Pagano RE (2007) Pathways of clathrin-independent endocytosis. Nat Rev Mol Cell Biol 8(8):603–612
19. Owens DE, Peppas NA (2006) Opsonization, biodistribution, and pharmacokinetics of polymeric nanoparticles. Int J Pharm 307(1):93–102
20. Kettler K et al (2014) Cellular uptake of nanoparticles as determined by particle properties, experimental conditions, and cell type. Environ Toxicol Chem 33(3):481–492
21. Champion JA, Mitragotri S (2009) Shape induced inhibition of phagocytosis of polymer particles. Pharm Res 26(1):244–249
22. Champion JA, Mitragotri S (2006) Role of target geometry in phagocytosis. Proc Natl Acad Sci U S A 103(13):4930–4934
23. Bareford LM, Swaan PW (2007) Endocytic mechanisms for targeted drug delivery. Adv Drug Deliv Rev 59(8):748–758
24. Yameen B et al (2014) Insight into nanoparticle cellular uptake and intracellular targeting. J Control Release 190:485–499
25. Nune SK et al (2009) Nanoparticles for biomedical imaging. Expert Opin Drug Deliv 6 (11):1175–1194
26. Howes MT et al (2010) Clathrin-independent carriers form a high capacity endocytic sorting system at the leading edge of migrating cells. J Cell Biol 190(4):675–691
27. Sahay G, Alakhova DY, Kabanov AV (2010) Endocytosis of nanomedicines. J Control Release 145(3):182–195
28. Harush-Frenkel O et al (2007) Targeting of nanoparticles to the clathrin-mediated endocytic pathway. Biochem Biophys Res Commun 353(1):26–32
29. Thomas CE, Ehrhardt A, Kay MA (2003) Progress and problems with the use of viral vectors for gene therapy. Nat Rev Genet 4(5):346–358
30. Misra SK et al (2015) Next generation carbon nanoparticles for efficient gene therapy. Mol Pharm 12(2):375–385
31. Misra SK et al (2014) Nanoscopic poly-DNA-cleaver for breast cancer regression with induced oxidative damage. Mol Pharm 11(11):4218–4227
32. Xu S et al (2013) Targeting receptor-mediated endocytotic pathways with nanoparticles: rationale and advances. Adv Drug Deliv Rev 65(1):121–138

33. Chou LY, Ming K, Chan WC (2011) Strategies for the intracellular delivery of nanoparticles. Chem Soc Rev 40(1):233–245
34. Morille M et al (2008) Progress in developing cationic vectors for non-viral systemic gene therapy against cancer. Biomaterials 29(24):3477–3496
35. Misra SK et al (2014) A cationic cholesterol based nanocarrier for the delivery of p53-EGFP-C3 plasmid to cancer cells. Biomaterials 35(4):1334–1346
36. Oh KS et al (2010) Core/shell nanoparticles for pH-sensitive delivery of doxorubicin. J Nanosci Nanotechnol 10(10):6967–6971
37. Lungwitz U et al (2005) Polyethylenimine-based non-viral gene delivery systems. Eur J Pharm Biopharm 60(2):247–266
38. Ferreira DDS et al (2013) pH-sensitive liposomes for drug delivery in cancer treatment. Ther Deliv 4(9):1099–1123
39. Huang JG, Leshuk T, Gu FX (2011) Emerging nanomaterials for targeting subcellular organelles. Nano Today 6(5):478–492
40. Berg K et al (1999) Photochemical internalization: a novel technology for delivery of macromolecules into cytosol. Cancer Res 59(6):1180–1183
41. Rajendran L, Knölker H-J, Simons K (2010) Subcellular targeting strategies for drug design and delivery. Nat Rev Drug Discov 9(1):29–42
42. Pasparakis G et al (2014) Harnessing photochemical internalization with dual degradable nanoparticles for combinatorial photo-chemotherapy. Nat Commun 5
43. Rittner K et al (2002) New basic membrane-destabilizing peptides for plasmid-based gene delivery in vitro and in vivo. Mol Ther 5(2):104–114
44. Vaccaro L et al (2005) Plasticity of influenza haemagglutinin fusion peptides and their interaction with lipid bilayers. Biophys J 88(1):25–36
45. Endoh T, Ohtsuki T (2009) Cellular siRNA delivery using cell-penetrating peptides modified for endosomal escape. Adv Drug Deliv Rev 61(9):704–709
46. Stewart KM, Horton KL, Kelley SO (2008) Cell-penetrating peptides as delivery vehicles for biology and medicine. Org Biomol Chem 6(13):2242–2255
47. Deshayes S et al (2005) Cell-penetrating peptides: tools for intracellular delivery of therapeutics. Cell Mol Life Sci 62(16):1839–1849
48. Chen B et al (2008) Transmembrane delivery of the cell-penetrating peptide conjugated semiconductor quantum dots. Langmuir 24(20):11866–11871
49. Holowka EP et al (2007) Polyarginine segments in block copolypeptides drive both vesicular assembly and intracellular delivery. Nat Mater 6(1):52–57
50. Wender PA et al (2000) The design, synthesis, and evaluation of molecules that enable or enhance cellular uptake: peptoid molecular transporters. Proc Natl Acad Sci 97 (24):13003–13008
51. Copolovici DM et al (2014) Cell-penetrating peptides: design, synthesis, and applications. ACS Nano 8(3):1972–1994
52. Torchilin VP (2006) Recent approaches to intracellular delivery of drugs and DNA and organelle targeting. Annu Rev Biomed Eng 8:343–375
53. Berry C (2008) Intracellular delivery of nanoparticles via the HIV-1 tat peptide
54. Ciobanasu C, Siebrasse JP, Kubitscheck U (2010) Cell-penetrating HIV1 TAT peptides can generate pores in model membranes. Biophys J 99(1):153–162
55. Karanth H, Murthy R (2007) pH-sensitive liposomes—principle and application in cancer therapy. J Pharm Pharmacol 59(4):469–483
56. Turk MJ et al (2002) Characterization of a novel pH-sensitive peptide that enhances drug release from folate-targeted liposomes at endosomal pHs. Biochim Biophys Acta Biomembranes 1559(1):56–68
57. Adarsh S, Shah Viral A, Umesh U (2011) Organelle specific targeted drug delivery – a review. Int J Res Pharm Biomed Sci 2:895–912
58. Suh J et al (2007) PEGylation of nanoparticles improves their cytoplasmic transport. Int J Nanomedicine 2(4):735

59. Yamada Y, Harashima H (2008) Mitochondrial drug delivery systems for macromolecule and their therapeutic application to mitochondrial diseases. Adv Drug Deliv Rev 60 (13):1439–1462
60. Biswas S, Torchilin VP (2014) Nanopreparations for organelle-specific delivery in cancer. Adv Drug Deliv Rev 66:26–41
61. Yamada Y et al (2008) MITO-porter: a liposome-based carrier system for delivery of macromolecules into mitochondria via membrane fusion. Biochim Biophys Acta Biomembranes 1778(2):423–432
62. Gruber J et al (2013) Mitochondria-targeted antioxidants and metabolic modulators as pharmacological interventions to slow ageing. Biotechnol Adv 31(5):563–592
63. Kopeček J (2013) Polymer–drug conjugates: origins, progress to date and future directions. Adv Drug Deliv Rev 65(1):49–59
64. Breunig M, Bauer S, Göpferich A (2008) Polymers and nanoparticles: intelligent tools for intracellular targeting? Eur J Pharm Biopharm 68(1):112–128
65. Weis K (2003) Regulating access to the genome: nucleocytoplasmic transport throughout the cell cycle. Cell 112(4):441–451
66. Griesenbach U et al (2012) Assessment of the nuclear pore dilating agent trans-cyclohexane-1,2-diol in differentiated airway epithelium. J Gene Med 14(7):491–500
67. Mohr D et al (2009) Characterisation of the passive permeability barrier of nuclear pore complexes. EMBO J 28(17):2541–2553
68. Keminer O, Peters R (1999) Permeability of single nuclear pores. Biophys J 77(1):217–228
69. Ellenberg J et al (1997) Nuclear membrane dynamics and reassembly in living cells: targeting of an inner nuclear membrane protein in interphase and mitosis. J Cell Biol 138(6):1193–1206
70. Kalderon D et al (1984) A short amino acid sequence able to specify nuclear location. Cell 39 (3):499–509
71. Lee BJ et al (2006) Rules for nuclear localization sequence recognition by karyopherinβ2. Cell 126(3):543–558
72. Wang R, Brattain MG (2007) The maximal size of protein to diffuse through the nuclear pore is larger than 60kDa. FEBS Lett 581(17):3164–3170
73. Mathot F et al (2007) Passive diffusion of polymeric surfactants across lipid bilayers. J Control Release 120(1):79–87
74. Dingwall C, Laskey R (1992) The nuclear membrane. Science 258(5084):942–947
75. Sakhrani NM, Padh H (2013) Organelle targeting: third level of drug targeting. Drug Des Dev Ther 7:585

Top Med Chem (2016) 20: 121–130
DOI: 10.1007/7355_2015_95

Published online: 26 August 2015

Personalized Medicine: Where Do We Go from Here?

Dipanjan Pan

Abstract The past decade has seen a surplus of nanotechnology-based methodologies for "theranostic" application, sensors, and real-time monitoring of biological events, therapy, and image-guided precision drug delivery. While nanotechnology offers great promise to address some of the critical clinical challenges, the future of this technology toward personalized medicine will largely be predisposed by design principles for developing translatable, "safer" nanoplatforms in concert with imaging agents, therapy, and homing ligands.

Keywords Molecular imaging, Multimodal imaging, Theranostics, Therapy

Contents

Abbreviations

ECM Extracellular matrix
ENM Engineered nanomaterial

D. Pan (✉)
Department of Bioengineering, University of Illinois at Urbana Champaign, Urbana, IL, USA

Beckman Institute for Science and Technology, Urbana, IL, USA

Department of Materials Science and Engineering, University of Illinois at Urbana Champaign, Urbana, IL, USA

Carle Foundation Hospital, Urbana, IL, USA
e-mail: dipanjan@illinois.edu

EPR	Enhanced permeability retention
I.V.	Intravenous
MPS	Mononuclear phagocyte system
MS	Mass spectrometry
NIPAM	*N*-Isopropylacrylamide
NLS	Nuclear localization signal
NMR	Nuclear magnetic resonance
NP	Nanoparticle(s)
PC	Protein corona
PEG	Polyethylene glycol
PS	Polystyrene
RES	Reticuloendothelial system
SWCNT	Single-walled carbon nanotubes
Tf	Transferrin
TfR	Transferrin receptor

1 Personalized Medicine: Perspective and Promises

This issue of *Topics in Medicinal Chemistry* is dedicated to the topic of personalized, futuristic medicine with a particular emphasis on molecular imaging and therapeutics [1–4]. We also conversed the role of functional nanometer-sized agents in this highly interdisciplinary area of science. Over the last decade, this multidisciplinary area of research has spawned unlimited attention, demonstrating high imminent for clinical translational. Advances in chemistry, biology, and engineering have formed unique scenarios for cross-disciplinary work at the molecular and cellular level, enabling unparalleled potential in early detection and therapy of a disease [4–6]. The unprecedented potential of nanoparticles in imaging and drug delivery has been well proven [6–9]. Significant advancement has been made to develop defined nanostructures for performing multiple function, i.e., targeting, imaging, and therapy.

The concept of theranostics is relatively new and exploits the "multifunctional" nature of nanoparticles. The smart nanoparticles capable of performing "dual function" are known as theranostic (therapy and diagnostics) platforms. Biological barriers are overcome by smart chemistry and by the incorporation of homing agents, contrast materials, and drugs [10–12]. Nanoparticle size, shape, and morphology are other critical parameters for successful clinical translation. Size dictates the in vivo characteristics of these agents designed for homing specific biological receptors and links with its bio-distributive nature, tissue buildup, and cellular internalization [13–15]. Targeting or homing nanoparticles to a diseased tissue can be categorized into two types – passive and active. In passive targeting approaches, the small particles ($<$100 nm) are believed to be up taken by the tumor vasculature due to their leakiness. This phenomenon is commonly known as enhanced permeability and retention (EPR) effect of tumors. For active targeting strategies, highly specific homing ligands such as peptides, antibodies, nucleic acids, aptamers, etc., are used. Currently there are nearly 40 nanoplatforms being explored at different levels of clinical stages. The majority of these agents rely on passive targeting approaches. A few passively targeted approaches are showing early promise in clinical trials (Tables 1 and 2), and many others are in the pipeline (Table 3) [16, 17]. Although EPR can permit nanoparticle passage in certain cancer tissues (e.g., inflammatory sites), most diseased tissues are not intrinsically characterized by the remarkably leaky vasculatures. For these pathological conditions, accumulation of nanoparticles will require an active mechanism of homing. The reduction of uptake of nanoparticles by healthy tissues (also tissues abundant with phagocytic cells) will require novel design by taking into consideration their size, morphology, surface properties, functional characteristics, etc. However, the exclusive identification and transport of nanoparticles in preferred pathologic cells of interest will essentially rely on active ligand-permitted homing. Choice of a homing agent is dependent on multiple critical variables, including (1) identification of a receptor having required cell specificity, cell surface density, degree of internalization, and trafficking channel, (2) identification of an agent with full specificity for the biological receptor, and (3) selective decoration of the agent with or without a linker to stimulate maximal projection of the ligand from the surface of the particles. In most of the cases, developing "ultimate" theranostic platform for imaging and therapeutics will be reliant upon careful consideration of the physicochemical characteristics of the particles and the biology of the tissue of interest. Tweaking will be necessary to alter the properties of these agents from initial proof-of-concept in vitro, ex vivo, and in vivo studies to clinical stages (Fig. 1).

RNAis are small noncoding RNAs that regulate gene expression and show crucial features in cancer genetics. Advanced sensitive high-throughput technologies help to understand the molecular and genetic network in cancer cells and the role of antisense agents as highly specific inhibitors of the expression of target genes to modulate the response of cancer cells to drug therapies [18–21]. The unparalleled potential of RNAi (miRNAs, siRNAs, etc.) is well established in a laboratory setting. Their delivery is restricted by numerous blockades, e.g., poor cellular uptake, toxicity, immunogenicity, unclear bio-distributive characteristics and clearance, degradation by nucleases, elimination by phagocytic immune cells, and reduced endosomal release. Biophysical techniques are engaged for the facile transport of biologics by two main ways – using viral vectors and nonviral vectors. Virus particles have inherent abilities to penetrate into the cells for the facile passage of biologics. Nonviral vectors, on the other hand, get transported to the cell either by physical (electroporation, gene gun) or chemical (utilizing lipoplexes and polyplexes) resources. With viral vectors, high transfection efficiencies can be achieved; however, their inability to carry large nucleic acid fragments makes them incompatible for many biological needs. They also exhibit high potential for mutagenesis, induce host immune responses, etc. The field of nonviral vectors is ruled by complexes of nucleic acids with cationic lipids (lipoplexes) and cationic polymers (polyplexes). Nonviral vectors based on nanometer-sized particles have shown great potential due to their low immunogenicity; however, a relatively lower

Table 1 Examples of approved nanomedicine agents in the clinic

Product	Nanoplatform/agent	Indication	Status	Company
Doxil	PEGylated liposome/ doxorubicin hydrochloride	Ovarian cancer	Approved 11/17/1995 FDA507I8	Ortho Biotech (acquired by JNJ)
Myocet	Non-PEGylated liposomal doxorubicin nanomedicine	Metastatic breast cancer	Approved in Europe and Canada, in combination with cyclophosphamide	Sopherion Therapeutics, LLC in North America and Cephalon, Inc. in Europe
DaunoXome	Lipid encapsulation of daunorubicin	First-line treatment for patients with advanced HIV-associated Kaposi's sarcoma	Approved in the USA	Galen Ltd.
ThermoDox	Heat-activated liposomal encapsulation of doxorubicin	Breast cancer, primary liver cancer	Received Fast Track Designation, approval expected by 2013	Celsion
Abraxane	Nanoparticulate albumin/paclitaxel	Various cancers	Approved 1/7/2005 FDA21660	Celgene
Rexin-G	Targeting protein-tagged phospholipid/ microRNA-122	Sarcoma osteosarcoma, pancreatic cancer, and other solid tumors	Phase II/III (Fast Track Designation, Orphan Drug Status Acquired) in the USA fully approved in the Philippines	Epeius Biotechnologies Corp.
Oncaspar	PEGylated asparaginase	Acute lymphoblastic leukemia	Approved 24/06/2006	Enzon Pharmaceuticals, Inc.
Resovist	Iron oxide nanoparticles coated with carboxydextran	Liver/spleen lesion imaging	In 2001, approved for the European market	Bayer Schering Pharma AG
Feridex	Iron oxide nanoparticles coated with dextran	Lesion imaging	Approved by US-FDA in 1996	Berlex Laboratories
Endorem	Iron oxide nanoparticles coated with dextran	Liver/spleen lesion imaging	Approved in Europe	Guerbet

Table 2 Examples of promising nanomedicine agents advanced stages of clinical trials

Medicine	Indication	Particle type	Company	Phase
PDS0101	Human papillomavirus-caused cancers	Positively charged liposome filled with antigen	PDS Biotechnology	Approved to begin Phase I
Bind-014	Prostate cancer	Tumor-targeting polymer nanoparticle filled with docetaxel	Bind Therapeutics	Approved to begin Phase II
Cyt-6091	Solid tumors	Gold nanoparticle linked to tumor necrosis factor	CytImmune Sciences	Phase II
AuroLase	Head and neck cancer, solid tumors	Gold nanoshells with silica core	Nanospectra Biosciences	Phase I
ATI-1123	Solid tumors	Liposome filled with docetaxel	Azaya Therapeutics	Phase I complete
PNT2258	Non-Hodgkin's lymphoma and other cancers	Liposome filled with DNA interference fragment	Pronai Therapeutics	Phase II

NCL has done preclinical testing on six therapeutics now in clinical trails
Sources: Companies, NCL

transfection rate is a challenge that should be taken care of before their clinical translation [21–24]. Successful clinical translation of this nonviral-based gene therapy will necessitate significant amount of work in terms of rational design, safety, optimized binding and release, etc. It is critical to gain better understanding of the delivery approaches to help design translatable platforms. Ideally, these platforms must not generate immune responses and would not be confronted by their unsettled efficacy, toxicity, and poor specificity.

The delivery of biologics in vivo is usually hindered by biological barriers such as reticuloendothelial system (RES) clearance, poor target specificity, and low overall tissue/cell penetration. Although majority of next-generation gene therapy agents still remained in preclinical stage, some have successfully entered clinical phases. LNA-modified-anti-miR-122 (Santaris, SPC3649) has entered the Phase III clinical stage for the treatment of hepatitis C virus (HCV) in liver transplant patients. Mirna Therapeutics is developing liposomal miR-34 construct (MRX34) for primary liver cancer. This technology is in Phase 1 clinical trial. Mark Davis and co-workers at Caltech reported that siRNA may successfully engage the human RNAi machinery to diminish expression of the M2 subunit of ribonucleotide reductase at both the mRNA and protein levels. The design of agent was based on nanoparticles (~70 nm diameter) stabilized by adamantane (AD)-terminated polyethylene glycol (PEG) complexed with a cyclodextrin-based polymer (CDP). These nanoparticles were targeted to cancer cells expressing the TF receptor using human transferrin (TF) protein for patients with solid tumors (melanoma) [23].

Emerging trends in molecular imaging (MI) bring promises to recognize the components, progressions, and dynamics of a disease at a molecular level. MI unites new contrast agents with biomedical modalities to visually identify cellular events and portray specific molecular traces in vivo in disease progression. The

Table 3 Examples of nanomedicine agents in various stages of clinical trials

Product/ agent	Nanoplatform	Indication	Status	Company
Cyclosert	Cyclodextrin nanoparticles (cyclodextrin NP/SiRNA)	Solid tumors	Phase I	Insert Therapeutics (now Calando Pharmaceuticals)
CRLX101	Cyclodextrin NPs/camptothecin	Various cancers	Phase II	Cerulean Pharma
S-CKD602	PEGylated liposomal CKD602 (topoisomerase inhibitor)	Various cancers	Phase I/II	Alza Corporation
CPX-1	Liposomal irinotecan	Colorectal cancer	Phase II	Celator Pharmaceuticals
CPX-351	Liposomal cytarabine and daunorubicin	Acute myeloid leukemia	Phase I	Celator Pharmaceuticals
LE-SN38	Liposomal SN38	Colorectal cancer	Phase II	Neopharm
INGN-401	Liposomal/FUS1	Lung cancer	Phase I	Introgen
NC-6004	Polymeric nanoparticle (PEG-polyaspartate) formulation of cisplatin	Various cancers	Phase I	NanoCarrier Co.
NK-105	Polymeric nanoparticle (PEG-polyaspartate) formulation of paclitaxel	Various cancers	Phase II	Nippon Kayaku Co. Ltd.
NK-911	Polymeric nanoparticle (PEC-polyaspartate) formulation of doxorubicin	Various cancers	Phase I	Nippon Kayaku Co Ltd.
NK-012	Polymeric micelle of SN-38	Various cancers	Phase II	Nippon Kayaku Co. Ltd.
SP1049C	Glycoprotein of doxorubicin	Various cancers	Phase II	Supratek Pharma Inc.
SPI-077	PEGylated liposomal cisplatin	Head/neck and lung cancer	Phase II	Alza Corporation
ALN-VSP	Lipid nanoparticle formulation of siRNA	Liver cancer	Phase I	Alnylam Pharmaceuticals
OSI-7904L	Liposomal thymidylate synthase inhibitor	Various cancers	Phase II	OSI Pharmaceuticals
Combidex	Iron oxide	Tumor imaging	Phase III	Advanced Magnetics
Aurimune	Colloidal gold/TNF	Solid tumors	Phase II	CytImmune Sciences

(continued)

Table 3 (continued)

Product/ agent	Nanoplatform	Indication	Status	Company
SGT-53	Liposome Tf anti-body/p53 gene	Solid tumors	Phase I	SynerGene Therapeutics
BIND 014	PLGA/PLA NPs/docetaxel	Prostate cancer and others	Phase I	BIND Biosciences
AuroLase	Gold-coated silica NPs	Head and neck cancer	Phase I	Nanospectra Biosciences
Rexin-G	Targeting protein-tagged phospholipid/ microRNA-122	Sarcoma, osteosarcoma, pancreatic cancer, and other solid tumors	Phase II/III (Fast Track Designation, Orphan Drug Status Acquired) in the USA fully approved in the Philippines	Epeius Biotechnologies Corp.
ThermoDox	Heat-activated liposomal encapsulation of doxorubicin	Brest cancer, primary liver cancer	Approved for breast cancer; Phase III for primary liver cancer	Celsion
BIND-014	Polymeric nanoparticle formulation of docetaxel	Various cancers	Phase I	BIND Bioscience
SGT53-01	Transferrin-targeted liposome with p53 gene	Solid tumors	Phase I	SynerGene Therapeutics
PEG-PGA and DON	PEG-glutaminase combined with glutamine antimetabolite 6-diazo-5-oxo-L-norleucine (DON)	Various cancers	Phase I/II	EvaluatePharma
PEG-IFNα2a	PEG-asys	Melanoma, chromic myeloid leukemia, and renal-cell carcinoma melanoma, multiple	Phase I/II	Genentech
ADI-PEG20	PEG-arginine deiminase	Hepatocellular carcinoma	Phase I	Polaris

recognition of the existing prospect to detect preclinical pathology has seen countless advancement in this area to synergistic development of sensitive, high-resolution imaging modalities, and molecular probes. The most frequently used noninvasive cellular and molecular imaging techniques include clinical modalities,

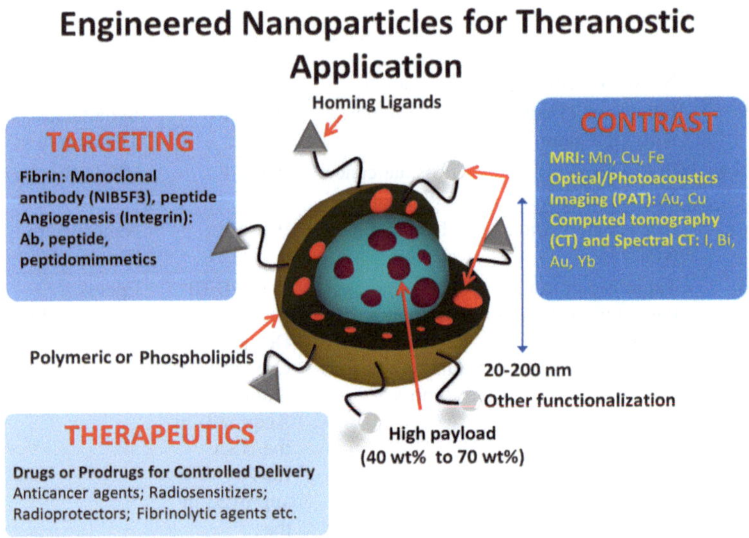

Fig. 1 Graphical representation of multifunctional nanoparticle with capabilities of running several tasks concurrently or exclusively

i.e., ultrasound (US), positron emission tomography (PET), computed tomography (CT), and magnetic resonance imaging (MRI). Some of the techniques are still preclinical in nature. These include optical, photoacoustic imaging (PAI), etc. Nanometer-sized agents for MI can be designed from various precursor materials (e.g., lipids, polymers, metals, etc.). They can act as a carrier to encapsulate a wide range of active constituents, including contrast agents and homing functionalities. Soft nanomaterials are usually derived from polymers, lipids, etc., and are outstanding examples for their flexibility for high payload and deliver to the disease site. Some of the materials are also known to respond to environmental factors (e.g., physiological or external stimuli). The physiological factors include pH, enzymatic, oxidative, and reductive conditions. The external stimuli are temperature, UV–Vis light, near-IR, stimulation with magnetic fields, or ultrasonic vibrations, etc.

2 Conclusion

The past decade has seen an excess of these approaches for theranostic application. The capability to monitor bio-signatures for early and noninvasive detection of a disease in permutation with directed therapy is the basis for nanomedicine. This interdisciplinary field is evolving rapidly and presenting clinically relevant promises as the science of molecular biology; genomics, chemistry, and nanotechnology advance significantly. However, with the increasing concern about the ethical and toxicity issues associated with some nano- "platforms," the biomedical researchers are in search of safer, more precise, and active way to bring nanomedicine to clinic. While nanotechnology offers boundless potential to address some of the burning issues in clinics today, its future in personalized medicine will largely be prejudiced by clever design philosophies for developing translatable, "safer" agents and by recognizing novel receptors and high-affinity homing agents.

References

1. Shin SJ, Beech JR, Kelly KA (2012) Targeted nanoparticles in imaging: paving the way for personalized medicine in the battle against cancer. Integr Biol (Camb) 5(1):29–42
2. Pang T (2012) Theranostics, the 21st century bioeconomy and 'one health'. Expert Rev Mol Diagn 12(8):807–809
3. Lee DY, Li KC (2011) Molecular theranostics: a primer for the imaging professional. AJR Am J Roentgenol 197(2):318–324
4. James ML, Gambhir SS (2012) A molecular imaging primer: modalities, imaging agents, and applications. Physiol Rev 92(2):897–965
5. Pan D, Lanza GM, Wickline SA, Caruthers SD (2009) Nanomedicine: perspective and promises with ligand-directed molecular imaging. Eur J Radiol 70(2):274–285
6. Alberti C (2012) From molecular imaging in preclinical/clinical oncology to theranostic applications in targeted tumor therapy. Eur Rev Med Pharmacol Sci 16(14):1925–1933
7. Wang LS, Chuang MC, Ho JA (2012) Nanotheranostics – a review of recent publications. Int J Nanomedicine 7:4679–4695
8. Lammers T, Aime S, Hennink WE, Storm G, Kiessling F (2011) Theranostic nanomedicine. Acc Chem Res 44(10):1029–1038
9. Cabral H, Nishiyama N, Kataoka K (2011) Supramolecular nanodevices: from design validation to theranostic nanomedicine. Acc Chem Res 44(10):999–1008
10. Kunjachan S, Jayapaul J, Mertens ME, Storm G, Kiessling F, Lammers T (2012) Theranostic systems and strategies for monitoring nanomedicine-mediated drug targeting. Curr Pharm Biotechnol 13(4):609–622
11. Pan D, Caruthers SD, Chen J, Winter PM, SenPan A, Schmieder AH, Wickline SA, Lanza GM (2010) Nanomedicine strategies for molecular targets with MRI and optical imaging. Future Med Chem 2(3):471–490
12. Lanza GM (2012) ICAM-1 and nanomedicine: nature's doorway to the extravascular tissue realm. Arterioscler Thromb Vasc Biol 32(5):1070–1071
13. Pan D, Pramanik M, Senpan A, Ghosh S, Wickline SA, Wang LV, Lanza GM (2010) Near infrared photoacoustic detection of sentinel lymph nodes with gold nanobeacons. Biomaterials 31(14):4088–4093
14. Albanese A, Tang PS, Chan WC (2012) The effect of nanoparticle size, shape, and surface chemistry on biological systems. Annu Rev Biomed Eng 14:1–16
15. Elsabahy M, Wooley KL (2012) Design of polymeric nanoparticles for biomedical delivery applications. Chem Soc Rev 41(7):2545–2561
16. Zhang L, Gu FX, Chan JM, Wang AZ, Langer RS, Farokhzad OC (2008) Nanoparticles in medicine: therapeutic applications and developments. Clin Pharmacol Ther 83(5):761–769
17. Wagner V, Dullaart A, Bock AK, Zweck A (2006) The emerging nanomedicine landscape. Nat Biotechnol 24(10):1211–1217
18. Bader AG, Brown D, Stoudemire J, Lammers P (2011) Developing therapeutic microRNAs for cancer. Gene Ther 18:1121–1126
19. Wang H, Jiang Y, Peng H, Chen Y, Zhu P, Huang Y (2015) Recent progress in microRNA delivery for cancer therapy by non-viral synthetic vectors. Adv Drug Deliv Rev 81C:142–160

20. Esau CC, Monia BP (2007) Therapeutic potential for microRNAs. Adv Drug Deliv Rev 59:101–114
21. Guo X, Huang L (2012) Recent advances in nonviral vectors for gene delivery. Acc Chem Res 45(7):971–979
22. Koynova R, Tenchov B (2011) Recent patents in cationic lipid carriers for delivery of nucleic acids. Recent Pat DNA Gene Seq 5(1):8–27
23. Davis ME, Zuckerman JE, Choi CH, Seligson D, Tolcher A, Alabi CA, Yen Y, Heidel JD, Ribas A (2010) Evidence of RNAi in humans from systemically administered siRNA via targeted nanoparticles. Nature 464(7291):1067–1070
24. Midoux P, Pichon C, Yaouanc JJ, Jaffrès PA (2009) Chemical vectors for gene delivery: a current review on polymers, peptides and lipids containing histidine or imidazole as nucleic acids carriers. Br J Pharmacol 157(2):166–178

Index

MIX
Papier aus verantwortungsvollen Quellen
Paper from responsible sources
FSC® C105338

If you have any concerns about our products,
you can contact us on
ProductSafety@springernature.com

In case Publisher is established outside the EU,
the EU authorized representative is:
Springer Nature Customer Service Center GmbH
Europaplatz 3, 69115 Heidelberg, Germany

Printed by Libri Plureos GmbH
in Hamburg, Germany